Progress
in
Pesticide Biochemistry

Volume 2

Progress
in
Pesticide Biochemistry

Volume 2

Edited by

D. H. Hutson

and

T. R. Roberts
Shell Research Ltd., Sittingbourne

A Wiley–Interscience Publication

JOHN WILEY & SONS
CHICHESTER · NEW YORK · BRISBANE · TORONTO · SINGAPORE

Library of Congress Cataloging in Publication Data (Revised):

Main entry under title:
Progress in pesticide biochemistry.
  Includes index.
  1. Pesticides—Physiological effect. 2. Biological
chemistry. I. Hutson, David Herd, 1935–   .
II. Roberts, Terence Robert, 1943–   .
QP82.2.P4P76   574.2′4   80–41419

ISBN 0 471 10118 4 (v. 2)   AACR2

British Library Cataloguing in Publication Data:

Progress in pesticide biochemistry.—Vol. 2
  1. Pesticides—Physiological effect—Periodicals
  2. Biological chemistry—Periodicals
  574.1′92   QP82.2P4

ISBN 0 471 10118 4

Set in Monophoto Times by
COMPOSITION HOUSE LTD., SALISBURY
Printed by
PAGE BROS. (NORWICH) LTD.

Contributors to Volume 2

V. T. Edwards   *Shell Research Ltd., Sittingbourne Research Centre, Sittingbourne, Kent, ME9 8AG, UK*

P. Hendley   *I.C.I. Plant Protection Division, Jealott's Hill Research Station, Bracknell, Berkshire, RG12 6EY, UK*

K. R. Huckle   *Department of Biochemistry, St. Mary's Hospital Medical School, Paddington, London, W2 1PG, UK*

D. H. Hutson   *Shell Research Ltd., Sittingbourne Research Centre, Sittingbourne, Kent, ME9 8AG, UK*

A. L. McMinn   *Shell Research Ltd., Sittingbourne Research Centre, Sittingbourne, Kent, ME9 8AG, UK*

P. Millburn   *Department of Biochemistry, St. Mary's Hospital Medical School, Paddington, London, W2 1PG, UK*

G. D. Paulson   *Metabolism and Radiation Research Laboratory, P.O. Box 5674, State University Station, Fargo, North Dakota 58105, USA*

L. O. Ruzo   *Laboratory of Pesticide Chemistry and Toxicology, Division of Entomology, University of California, Berkeley, California 94720, USA*

A. N. Wright   *Shell Research Ltd., Sittingbourne Research Centre, Sittingbourne, Kent, ME9 8AG, UK*

Contents

vii

Preface

At a time when the traditional screening process of discovery is being increasingly complemented by the biochemical approach, an understanding of the biochemistry and mode of action of insecticides, fungicides, herbicides and growth regulators is fundamental for both design of new compounds and their safety evaluation. Primary publications on the biochemistry of pesticides are distributed widely throughout the scientific literature and the subject matter ranges from insect, plant and soil biochemistry through to mammalian toxicology. This series is one in which selected aspects will be reviewed, and interrelated where possible.

Within the scope of the series, the biochemistry of pesticides is deemed to cover the following areas: (1) mode of action (i.e. biocidal action in target species); (2) biotransformation in target species (which may, of course, be related to mode of action); (3) biotransformation in non-target species (which may include soils, bacteria, insects, plants, fish, birds and mammals, including man); (4) environmental effects (e.g. effects on the ecology of treated areas); (5) environmental chemistry (distribution and fate in the environment); and (6) biochemical toxicology in mammals, including man. In selecting areas for review, we take account of new or changing aspects of the science, including techniques, and of changes in the importance or use patterns of the chemical classes of pesticides. This is exemplified in the subjects covered in Volume 1, the contents of which are listed at the end of this volume.

Conjugation of pesticides and their primary metabolites is an important metabolic process in plants and animals and has received considerable attention as a topic of research in recent years. In planning Volume 2 of *Progress in Pesticide Biochemistry*, it was our aim to include some of these recent developments. Conjugates are generally regarded as detoxification products of the bioactive molecule. That this is predominantly the case is not surprising considering the drastic change in physical properties which occurs on the conversion of a generally lipophilic compound to a polar, ionized conjugate. Whilst the generalization is true, as in other aspects of xenobiochemistry, the great variety of chemical structures encountered ensures that exceptions occur. Thus, the simple sugar conjugates are apparently not necessarily terminal metabolites in plants but may be hydrolysed to reform a biologically active compound, converted into di- and oligo-saccharides or esterified with malonic acid (Chapter 3). Amino acid conjugates of herbicidal acids sometimes either possess intrinsic herbicidal activity or also act as storage forms of the herbicide. Conjugation processes are also known to play a role in the toxicity of aromatic amines and other xenobiotics (Chapter 6). The generalization that all conjugates are readily excreted polar molecules now has its exception with the discovery in the last few years of a number of xenobiotic lipids, which are biosynthesized via the established conjugation mechanisms but involving unusual acceptors (Chapter 5). A fourth chapter on conjugation processes has been included to summarize the reasons behind the sometimes remarkable species differences encountered in amino acid conjugation (Chapter 4).

Continuing our inclusion in this series of developments in currently important techniques, the use of stable isotopes in metabolic, analytical and mechanistic studies with pesticides is reviewed in Chapter 2. A review of the metabolism of the pyrethroid insecticides in plants and soils was included in Volume 1 of *Progress in Pesticide Biochemistry* and some aspects of the fate and impact of this class of compounds in the environment will be included in subsequent volumes. In this volume current information on the photochemistry of the pyrethroids is summarized (Chapter 1). Although photochemistry is not itself a biochemical process it may have a profound effect on the interaction of a pesticide with target and non-target species (biochemical processes) and therefore merits consideration.

Sittingbourne, 1981   D. H. HUTSON
                                                           T. R. ROBERTS

Progress in Pesticide Biochemistry, Volume 2
Edited by D. H. Hutson and T. R. Roberts
© 1982 John Wiley & Sons, Ltd.

CHAPTER   1

Photochemical reactions of the synthetic pyrethroids

L. O. Ruzo

INTRODUCTION

The photochemical breakdown of chemicals in the environment is a subject that merits discussion since it is in many cases the primary pathway for disposal of these chemicals. One of the most important aspects concerning insecticides is to determine their effect on the environment as well as on non-target organisms. Much of what occurs after the application of a pesticide can be determined by its photoreactions mediated by sunlight. This is especially true of the pyrethroids, which contain several light-absorbing moieties and can thus be expected to yield considerable numbers of photoproducts. Another important consideration which arises from photochemical studies includes determination of photoproduct toxicities and persistence. Several photoprocesses undergone by pyrethroids closely resemble their metabolic reactions, if not mechanistically, at least as far as the identity of the resulting products, i.e. as in ester cleavage and oxidation reactions. Furthermore, much of the degradation of pyrethroids on plants is due to their photochemical processes. The study of pesticide biochemistry

1

must take into account the variety of products which are unintentionally introduced into the ecosystem since ultimately these will be partially responsible for interactions with various organisms.

The field of pesticide photochemistry has only developed in recent years; this applies especially to the pyrethroids. In this review an attempt has been made to include all the published work on the subject up to November 1980. It is the author's hope that it will serve as a useful summary and emphasize the importance of the processes described for successful utilization of these potent insect control chemicals.

BASIC PROCESSES IN THE PHOTOCHEMISTRY OF PESTICIDES

Two separate types of energy absorption are of interest in considering the photoreactions of any organic molecule. Direct absorption of a photon by a molecule results in an increase in energy of the electrons at the site of absorption (chromophore). This variation in the energy does not initially result in a change of the spin orientation of the electrons. Since a spin change is generally forbidden, the initial excited state retains the same spin multiplicity of the ground state, for most organic compounds this is one in which all the electrons are paired. This species is known as a singlet (S). Once the molecule is in the excited state; a variety of factors may bring about spin-orbit coupling, a process that breaks down, at least partially, the forbidden character of intersystem crossing (isc) resulting in the formation of an excited state in which electron spins are unpaired. Because of a spin multiplicity value of three this excited state is a triplet (T). The triplet excited state is usually of lower energy and longer lifetime than the singlet since its return to the singlet ground state is also forbidden by spin multiplicity rules.

The second mode of energy absorption may occur when another type of molecule is present in the chemical environment of the substrate of interest and can donate the energy it absorbs to the latter. The energy donor (D) is known as the sensitizer and the acceptor (A) as the quencher. Because of lifetime considerations sensitizers are usually triplets and thus provide a route for direct population of the substrate's triplet manifold without intermediacy of the singlet state.

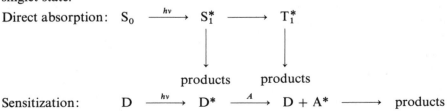

Much of the absorbed radiation may be re-emitted by S_1 or T_1 as fluorescence and phosphorescence respectively or it may be dissipated by collisions in a radiationless process. The efficiency of the reactions leading to product formation is most conveniently measured by determination of the quantum yield (ϕ) which represents the number of substrate molecules that react per photon of radiation absorbed. The value of ϕ is generally obtained by simultaneously irradiating the compound of interest and one whose quantum yield is already known (an actinometer) with monochromatic light.

The photodecomposition of a pesticide requires two events: absorption of radiation (direct or sensitized) and transformation of the electronically excited species by chemical processes. The first chemical step in a photoreaction involves bond cleavage, usually homolytic, which yields free radical species. The energy equivalent in radiation of environmentally significant wavelengths is sufficient to disrupt most covalent bonds, i.e. 300 nm corresponds to 95 kcal. mole^{-1}. Environmental photodegradation must occur at wavelengths above 290 nm since the ozone layer absorbs most of the sun's electromagnetic radiation below that wavelength. Many pesticides, especially among the chlorinated hydrocarbons have their major absorption bands below 290 nm and are thus photodecomposed slowly. However, this is not the case with pyrethroids since they contain a variety of chromophores whose combined effect results in appreciable extinction coefficients, and therefore absorbance, at sunlight wavelengths. The number of chromophores in a pyrethroid molecule does not allow for these compounds to be grouped together as a class from a photochemical standpoint. Fortunately, enough similarities exist so that certain types of reactions are common to many pyrethroids. These reactions will be emphasized in the present review.

The pyrethrins and earlier synthetic pyrethroids have been known to lack photostability for many years (Crowe et al., 1961), but this aspect of their chemistry was viewed mainly as a drawback to their use in the field. More recently, with the advent of potent and photostable pyrethroids (Elliott, 1977), questions have arisen regarding the identity, toxicology and persistence of the resulting photoproducts. The photochemistry of pyrethroids has been reviewed in part by Holmstead et al. (1976) and by Ruzo (1980). A useful review on pesticide photochemistry is that of Zabik et al. (1976). The structures of the pyrethrins and pyrethroids discussed in this review are shown in Figure 1.

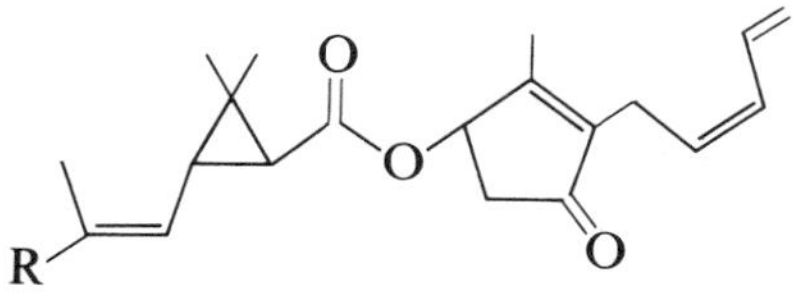

Pyrethrin I, R = CH$_3$
II, R = CO$_2$CH$_3$

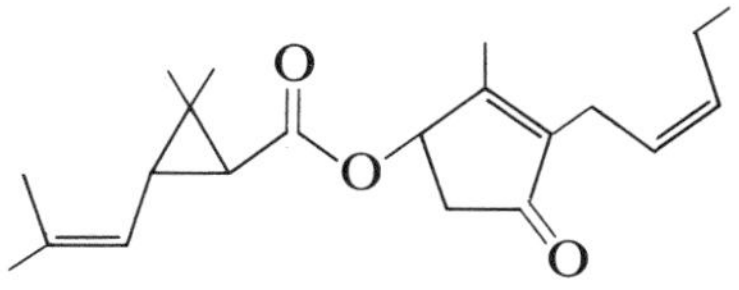

Jasmolin I

Figure 1 (*continued*)

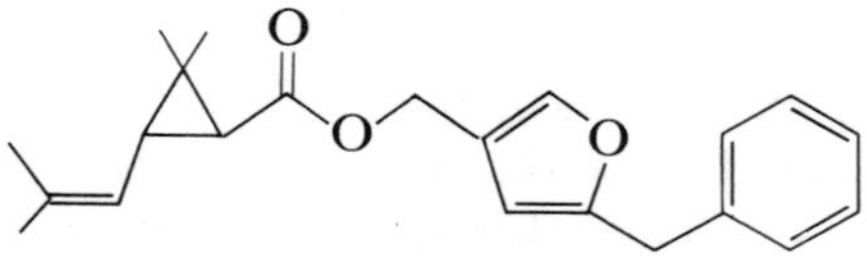

t-Resmethrin

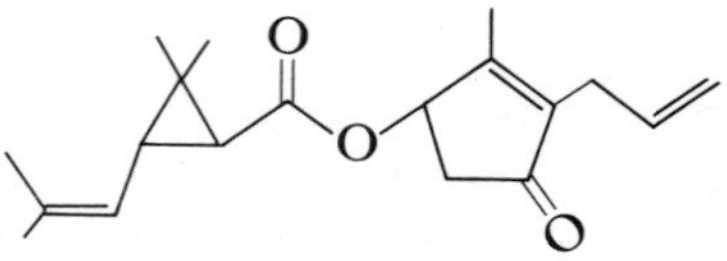

t-Allethrin

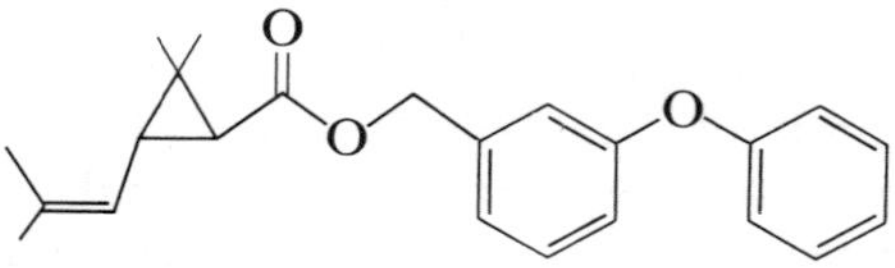

t-Phenothrin

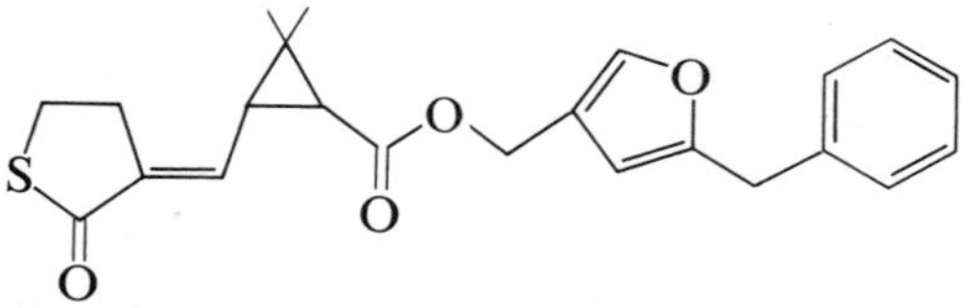

Kadethrin

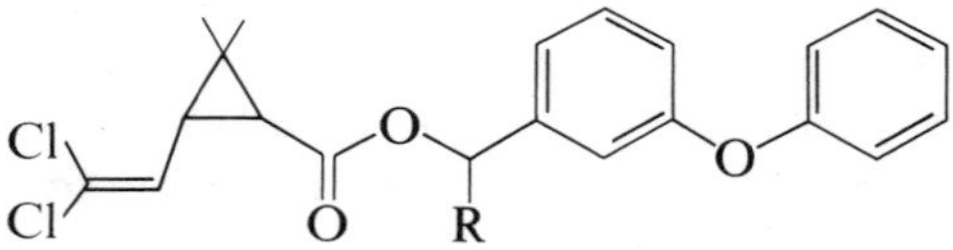

c-Permethrin, R = H
c-Cypermethrin, R = CN

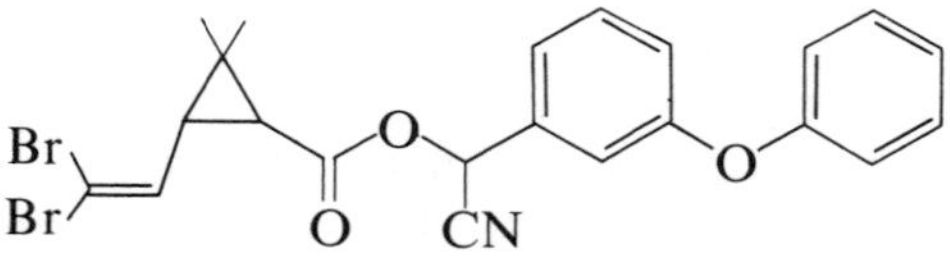

Decamethrin (deltamethrin)

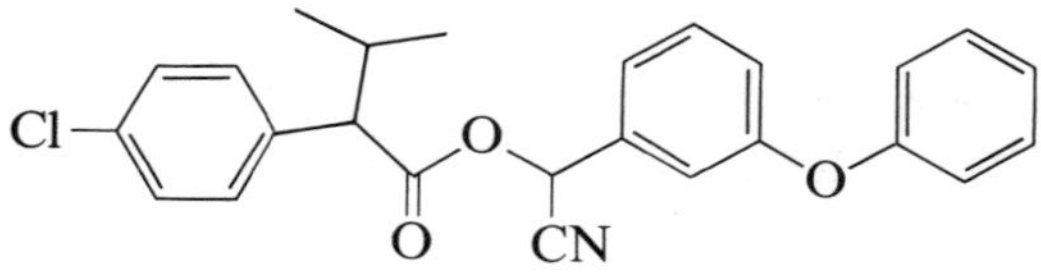

Fenvalerate

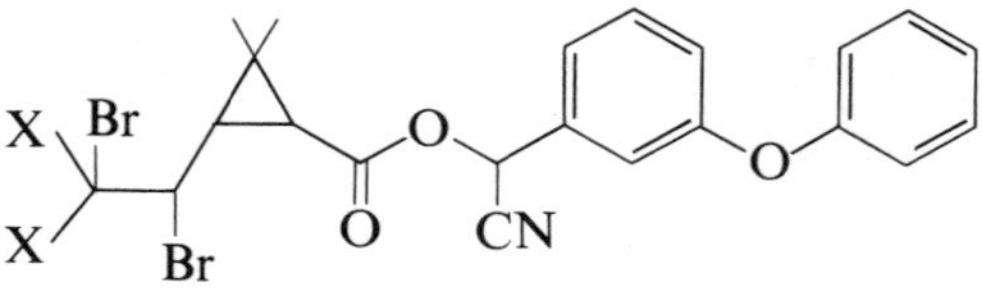

Tralomethrin, X = Br
Tralocythrin, X = Cl

Figure 1 Structures of some natural and synthetic pyrethroids

EARLY PHOTOCHEMICAL STUDIES OF THE NATURAL PYRETHRINS AND THEIR DERIVATIVES

The insecticidal components of pyrethrum extract are rapidly photodecomposed by sunlight to inactive materials. Somewhat contradictory findings were reported in the early stages of investigation, the acid and alcohol portions were alternately believed to be the primary site responsible for photodecomposition (Brooke, 1967; Glynne Jones, 1960; Head et al., 1968). These apparent contradictions probably arose from differences in purity of the irradiated materials, irradiation conditions and light sources employed (Chen and Casida, 1969).

Exposure of purified pyrethrum extract in petroleum ether to sunlight for three months produced resinous materials which upon hydrolysis yielded

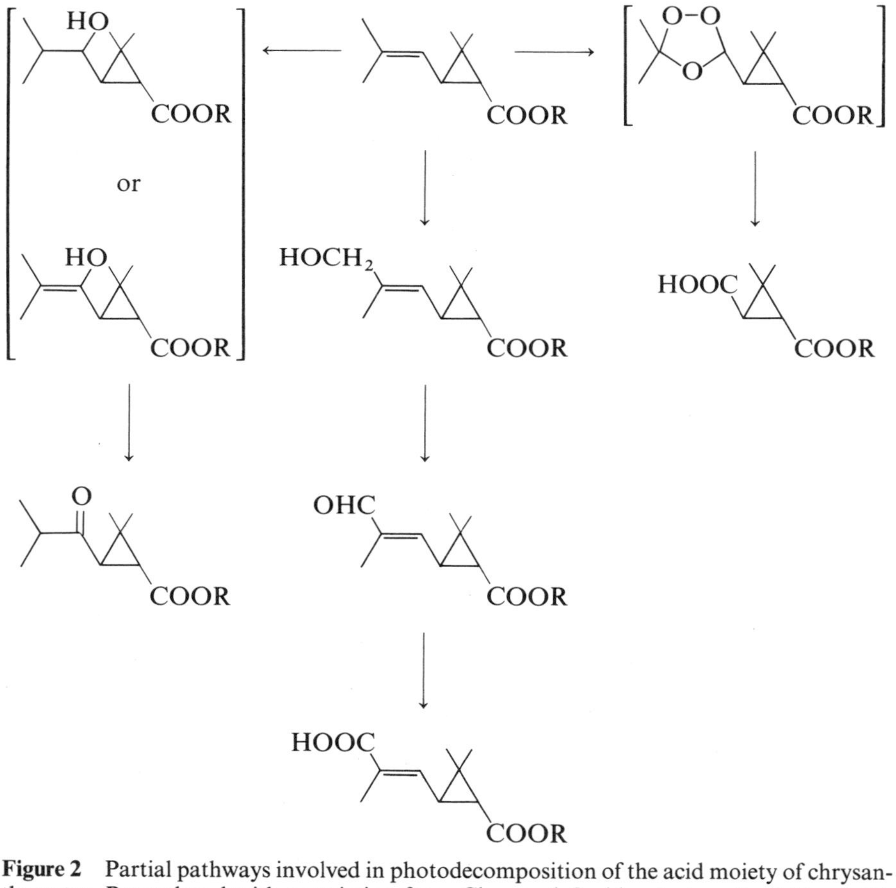

Figure 2 Partial pathways involved in photodecomposition of the acid moiety of chrysanthemates. Reproduced with permission from Chen and Casida, *J. Agr. Food Chem.*, **17**, 208–215, © 1969, American Chemical Society

chrysanthemic and chrysanthemumic dicarboxylic acids (Campbell and Mitchell, 1950). Freeman (1956) suggested that pyrethrins were photode-composed faster than the cinerins and that allethrin was decomposed at an even slower rate. These findings were confirmed by Brown *et al.* (1957). Their studies also indicated that changes occurred only in the cyclopentenolone portion of the molecule. Stahl (1960) demonstrated that when photolysed on silica gel the order of reactivity is pyrethrin II > pyrethrin I > cinerin I > cinerin II. Each of these compounds yields a material which is sensitive to peroxide reagents and the photoproduct mixture is of greatly reduced insecticidal activity.

The decomposition of pyrethrin I, allethrin, tetramethrin and dimethrin as thin films irradiated with a sunlamp yielded products which upon saponification allowed for identification of some of the acid moiety metabolites (Figure 2) (Chen and Casida, 1969).

PHOTOREACTIVITY OF PYRETHROIDS

The pyrethroids that are derivatives of chrysanthemic acid contain an isobutenyl group which is very susceptible to biological (Yamamoto and Casida, 1966; Yamamoto *et al.*, 1969) and photochemical oxidation (Ueda *et al.*, 1974; Chen and Casida, 1969; Ruzo *et al.*, 1980). This photolabile moiety has been replaced with dihalogenovinyl substituents in the insecticidally potent compounds permethrin (Elliott *et al.*, 1973), cypermethrin (Burt *et al.*, 1974) and decamethrin (Elliott *et al.*, 1974a), and with 4-chlorophenylisovalerate in fenvalerate (Ohno *et al.*, 1974). These compounds exhibit two major ultraviolet absorption bands, a relatively intense one at 210–230 nm ($\varepsilon > 1000$) for the alcohol π–π^* transition of the unsaturated groups and another at 250–280 nm ($\varepsilon > 100$) corresponding to the carbonyls which is essentially n–π^* in character. It is the latter transition that is responsible for environmental photodegradation.

Several comparative studies of pyrethroid photolability have been reported, but in some cases, the results are conflicting. Resmethrin has been found to be more stable than allethrin and tetramethrin when irradiated as a thin film on glass (Abe *et al.*, 1972), but more reactive when absorbed on silica gel (Ueda *et al.*, 1974). The photoreactivity of the acid moiety of the potent knockdown pyrethroid kadethrin has been found to be greater than that of the analogous lactone-substituted chrysanthemate, the pyrethrate and of the chrysanthemate when the corresponding [d]2-octyl esters are sensitized by acetone or iso-butyrophenone (Ohsawa and Casida, 1979).

The reaction quantum yields of six pyrethroids and of the methyl dichloro- and dibromovinyl cyclopropane carboxylates (Ruzo and Casida, 1980) (Table 1) indicate a three- to ten-fold difference in reactivity between allethrin and the phenoxybenzyl compounds. Since the methanol solutions absorbed $>90\%$ of the incident light and the samples were irradiated simultaneously, the results are directly comparable. It is evident that much of the absorbed radiation in the phenoxybenzyl esters result in no net reaction since the corresponding methyl

Table 1 Reaction quantum yields of degradation of pyrethroids and related compounds at $\lambda > 290$ nm in methanol solution (reproduced from Ruzo and Casida, 1980, with permission of the Royal Society of Chemistry)

Compound	ϕ_{300}
S-Bioallethrin (1R,trans,4'S)	0.216
Fenothrin (1R,cis)	0.070
Permethrin (1RS,cis)	0.018
Cypermethrin (1RS,cis)	0.022
Decamethrin (1R,cis)	0.043
Fenvalerate (1RS,αRS)	0.043
Methyl cis-2,2-dimethyl-3-(2,2-dihalovinyl)-cyclopropane carboxylates:	
X = Cl	0.005
X = Br	0.121

esters exhibit a greater value of ϕ. This suggests that although photostability is improved by introduction of the dihalovinyl group, the molecule retains considerable reactivity in the absence of an intramolecular ultraviolet filter, i.e. the alcohol moiety.

The acid moiety of pyrethroid esters has been further modified by use of 3-alkoxycyclopropane carboxylates. These compounds are somewhat more stable photochemically than permethrin (Minemite et al., 1978) probably because they lack the vinyl chromophore.

Photochemical ester cleavage

In general three mechanisms can accomplish the rupture of ester bonds: scission of the carboxylate–carbon bond (Equation 1) yielding carboxyl and alkyl free radicals (the resulting RCOO· moiety can further decompose and give carbon dioxide); cleavage of the carbonyl–oxygen bond (Equation 2) and subsequent production of carbon monoxide; photonucleophillic attack by solvent or other nucleophiles (N) (Equation 3) at the excited ester carbonyl.

$$R-\overset{O}{\underset{OR_1}{\|}}\|\ \longrightarrow\ R-\overset{O}{\underset{O\cdot}{\|}}\|\ +\ R_1\cdot\ \longrightarrow\ R\cdot + CO_2 \qquad (1)$$

$$R-\overset{O}{\underset{OR_1}{\|}}\|\ \longrightarrow\ R-\overset{O}{\|}\|\ +\ R_1O\cdot\ \longrightarrow\ R\cdot + CO \qquad (2)$$

$$R-\overset{O}{\underset{OR_1}{\|}}\|\ \overset{}{\underset{N}{\longrightarrow}}\ R-\overset{O^-}{\underset{N}{|}}OR_1\ \longrightarrow\ R-\overset{O}{\underset{N}{\|}}\|\ +\ R_1O^- \qquad (3)$$

The primary products from pyrethroid cleavage are often unstable and under-go further photoreaction. These reactions are discussed in subsequent sections.

Cleavage of the cis- and trans-isomers of resmethrin upon exposure to sun-light wavelengths on silica gel or aqueous solution (Ueda et al., 1974) yields the corresponding chrysanthemic acid, but not 5-benzyl-3-furylmethanol. How-ever, ester cleavage is not an important reaction for pyrethrin I, allethrin, tetramethrin or dimethrin when irradiated as thin films on glass by sunlight wavelengths (Chen and Casida, 1969). A separate study identified chrysanthemic acid from S-bioallethrin under a variety of irradiation conditions (Ruzo et al., 1980), but the alcohol moiety photoproducts were not characterized. Kadethrin is cleaved in the solid phase by sunlight both at the ester and thiolactone groups (Ohsawa and Casida, 1979). The alcohol moiety was detected as the oxidized form in benzyl furylcarboxylic acid (Figure 3).

Photolysis of the ester bond is a major reaction in the newer pyrethroids such as decamethrin (Ruzo et al., 1976, 1977), permethrin (Holmstead et al., 1978a) and fenvalerate (Holmstead et al., 1978b). When trans-permethrin is irradiated in hexane or water solution ($>$ 290 nm) the isomeric dichlorovinyl acids and 3-phenoxybenzyl alcohol are obtained; in methanol the corresponding ester and ether are formed (Figure 4). cis-Permethrin undergoes analogous reactions. Photolysis of decamethrin in alcohols, hexane and aqueous solutions ($\lambda >$ 290 nm) results mainly in two modes of ester cleavage (Equations 1, 3) including decarboxylation as a minor pathway (Figure 5). The reaction rate decreases in the solvent order methanol > ethanol > 2-propanol either because of decreased nucleophilicity, minimizing Equation 3, or by increased viscosity allowing for greater recombination of the initially formed radicals in a solvent cage. Major products from the decamethrin acid moiety in methanol include the dibromo- and bromovinyl acids, their methyl esters and dibromovinyl-cyclopropane aldehyde. Trace amounts of either dibromovinylcyclopropane or its ring-opened isomer are formed, presumably by hydrogen abstraction by the radical formed upon decarboxylation. Photolysis of thin films of decamethrin by sunlight results in similar modes of ester cleavage in solution.

The alcohol moiety of decamethrin yields α-cyano-3-phenoxybenzyl alcohol, 3-phenoxybenzaldehyde and 3-phenoxybenzoyl cyanide, the latter compound is isolated only in hexane, while in methanol or aqueous solutions the corre-sponding methyl 3-phenoxybenzoate and 3-phenoxybenzoic acid are obtained. The cyanohydrin is thermally and photochemically unstable, eliminating HCN to yield the aldehyde.

Fenvalerate is rapidly photodecomposed in methanol, hexane or acetonitrile-water solutions upon irradiation at $\lambda >$ 290 nm. The major product ($>$ 60%) in all cases is decarboxylated material (Figure 6). Decarboxylation is also a major reaction of thin films on glass. Other ester cleavage products obtained from the acid moiety arise by reactions of the secondary benzyl radical formed upon extrusion of carbon dioxide. The photoproducts from the alcohol moiety resemble those obtained from decamethrin.

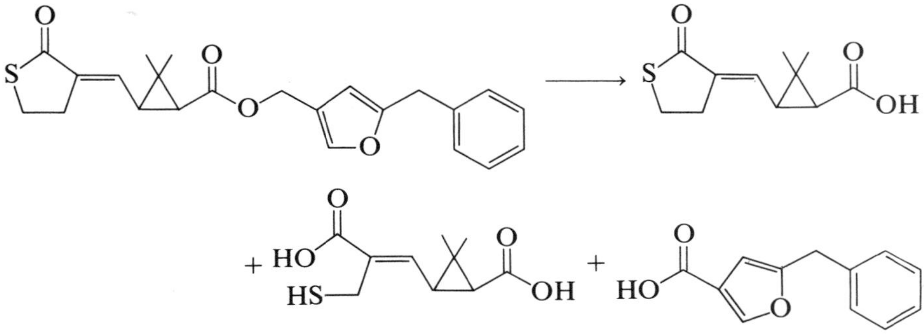

Figure 3 Photohydrolysis of kadethrin upon sunlight irradiation. Reproduced with permission from Ohsawa and Casida, *J. Agr. Food Chem.*, **27**, 1112–1120, © 1970, American Chemical Society

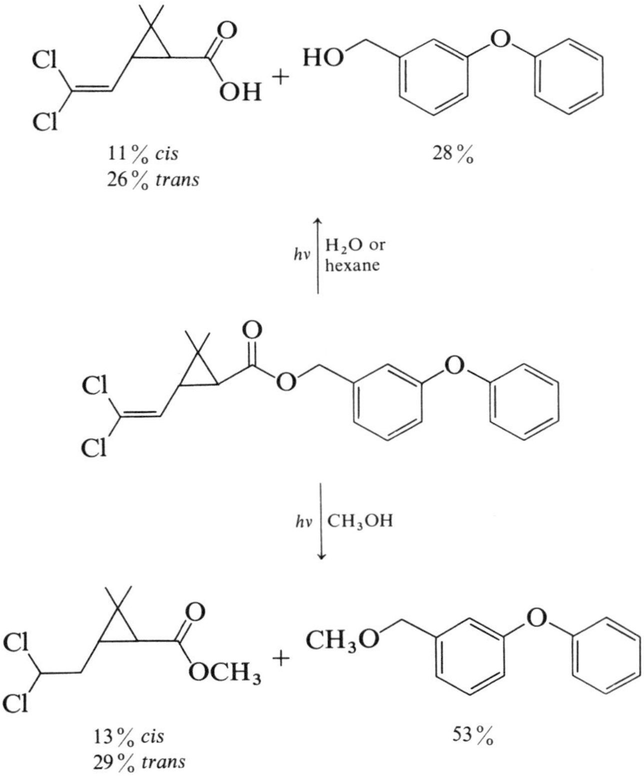

Figure 4 Ester cleavage reactions of *t*-permethrin. Reproduced with permission from Holmstead *et al.*, *J. Agr. Food Chem.*, **26**, 590–595, © 1978 American Chemical Society

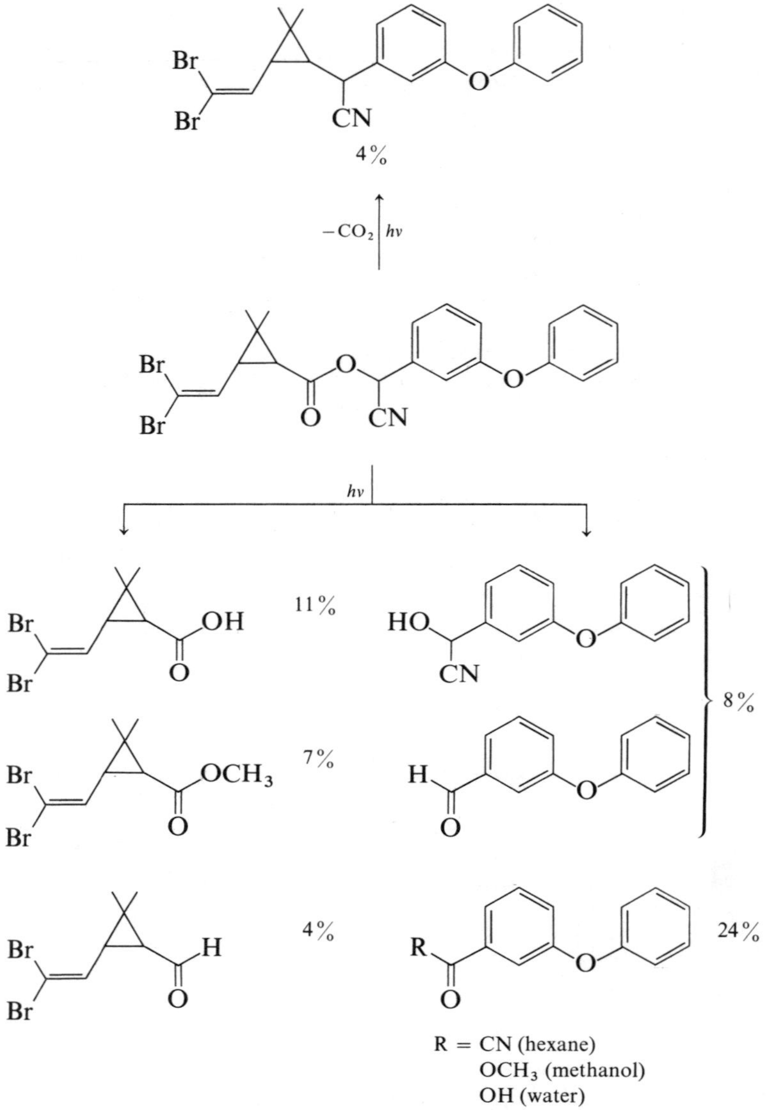

Figure 5 Ester cleavage reactions of decamethrin. Reproduced with permission from Ruzo *et al.*, *J. Agr. Food Chem.*, **25**, 1385–1394 © 1977 American Chemical Society

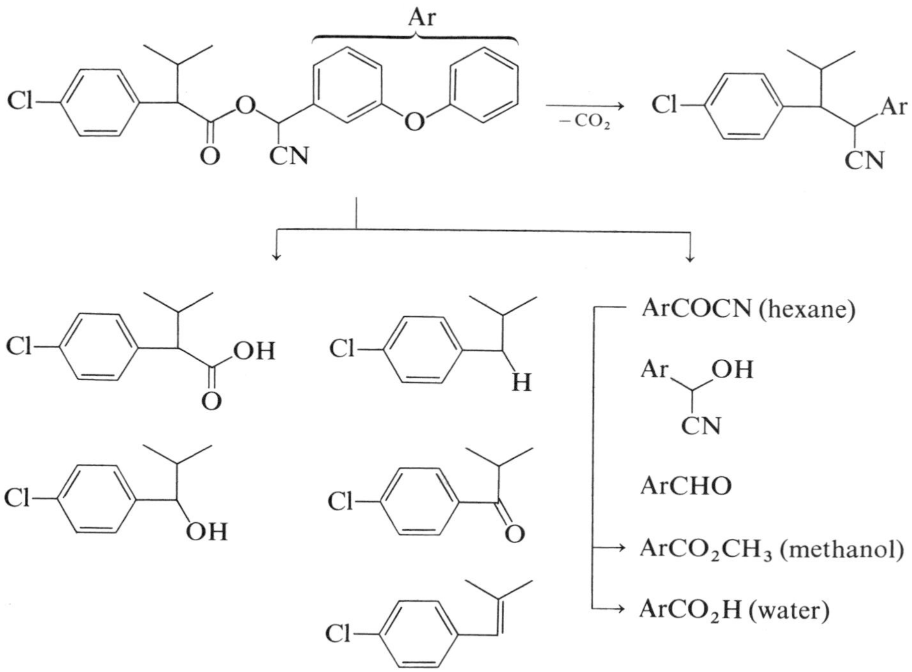

Figure 6 Ester cleavage reactions of fenvalerate. Reproduced with permission from Holmstead *et al.*, *J. Agr. Food Chem.*, **26**, 954–959 © 1978 American Chemical Society

One of the available pathways for photoreaction in esters involves loss of carbon dioxide. This is the major process undergone by several acetates in solution (Zimmerman and Sandel, 1963; Van Dusen and Hamill, 1962). The decarboxylation mechanism has been studied in benzyl acetate (Givens and Oettle, 1972) and naphthol esters (Matuszewski *et al.*, 1973) and the results suggest that photoelimination occurs from discrete radical intermediates in a stepwise manner from the triplet state with subsequent recombination of the radicals in the solvent cage. Several cyanohydrin esters with structures similar to fenvalerate have been found to decarboxylate readily (Holmstead and Fullmer, 1977). The increased importance of decarboxylation in fenvalerate ($\sim 60\%$) relative to decamethrin (4%) has been explained on the basis of stability of the resulting radicals (Ruzo *et al.*, 1977; Holmstead *et al.*, 1978b). Fenvalerate generates two benzylic radicals both of which are stabilized by substituents, while decamethrin would yield preferably a carboxyl radical which abstracts hydrogen to form the dihalovinyl cyclopropanecarboxylic acid. Photoelimination of CO_2 is negligible in unsubstituted benzyl esters such as permethrin (Holmstead *et al.*, 1978a).

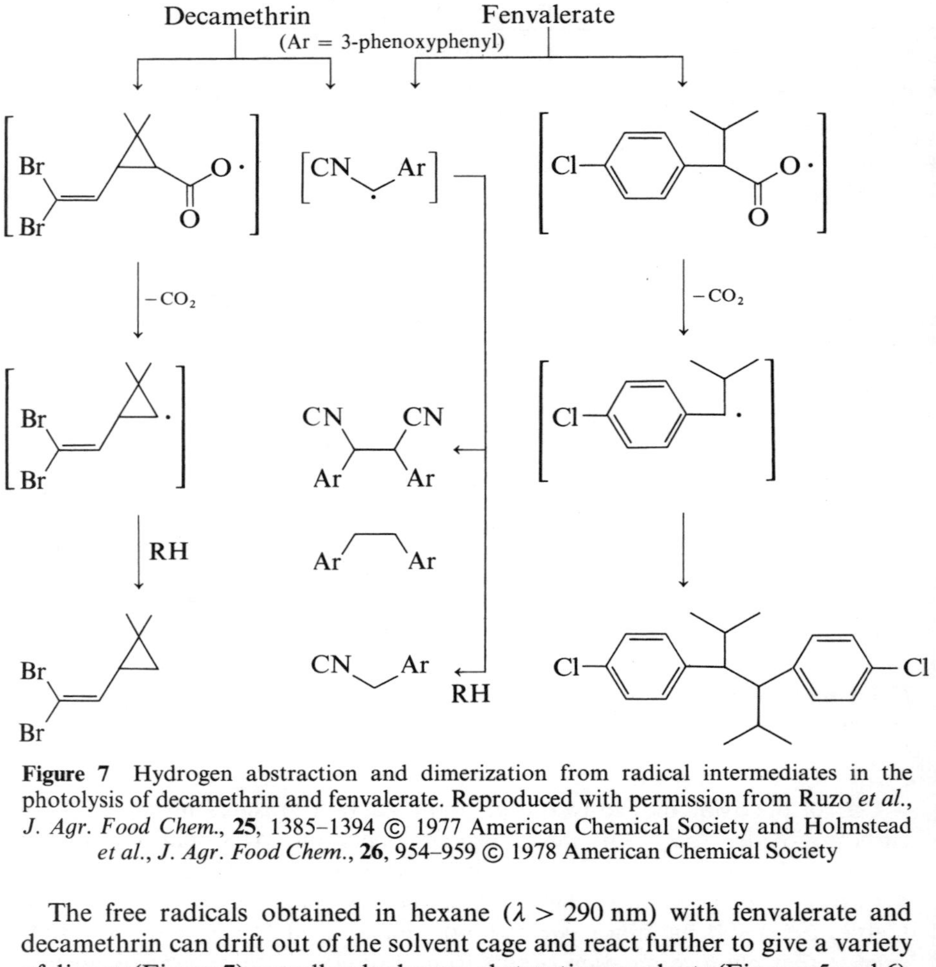

Figure 7 Hydrogen abstraction and dimerization from radical intermediates in the photolysis of decamethrin and fenvalerate. Reproduced with permission from Ruzo *et al.*, *J. Agr. Food Chem.*, **25**, 1385–1394 © 1977 American Chemical Society and Holmstead *et al.*, *J. Agr. Food Chem.*, **26**, 954–959 © 1978 American Chemical Society

The free radicals obtained in hexane ($\lambda > 290$ nm) with fenvalerate and decamethrin can drift out of the solvent cage and react further to give a variety of dimers (Figure 7) as well as hydrogen abstraction products (Figures 5 and 6).

Much greater yields of dimers from the alcohol moiety are obtained with decamethrin than with fenvalerate. Dimerization is favoured in degassed solution under ultraviolet irradiation, but not in the solid phase under sunlight.

Reactions involving the cyclopropyl group

The insecticidal activity of pyrethroids is largely dependent on the configuration at carbon 1 (adjacent to carboxyl) of the cyclopropane ring (Elliott *et al.*, 1974b). The most active compounds have a 1R configuration at this site. Consequently, epimerization to the 1S-isomer results in dramatic loss of activity. Furthermore, the *cis*- and *trans*-isomers (about C_1 and C_3) differ considerably

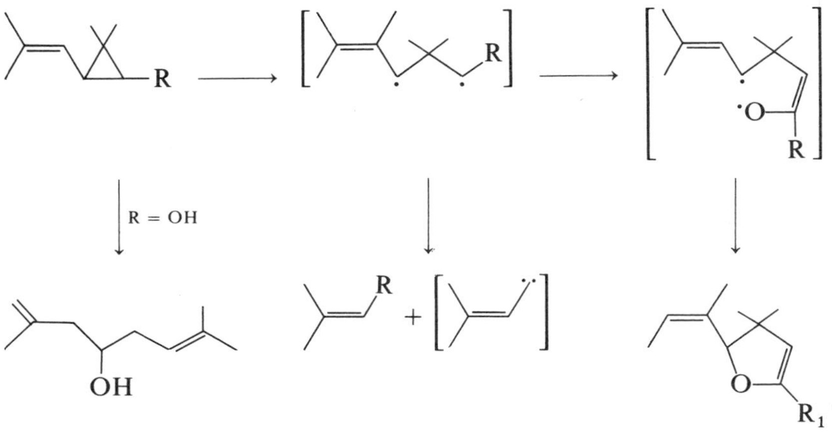

Figure 8   Photochemical rearrangement in the cyclopropyl moiety of chrysanthemic acid and its derivatives

in biostability and insecticidal activity (Elliott *et al.*, 1974a; Casida *et al.*, 1979). These factors make photoisomerization an important reaction of pyrethroids.

Much of the early work on *cis–trans* isomerization processes in pyrethroids was carried out on simple derivatives, such as the alkyl esters of chrysanthemic acid. Sasaki *et al.* (1970) reported the modifications in the cyclopropyl group of chrysanthemic acid and the corresponding alcohol, amide and ethyl ester in hexane solution using the unfiltered light from a high pressure mercury source. The major reactions were lactone formation and elimination of a carbene to yield dimethylacrylate (Figure 8). Chrysanthemol was photoisomerized to 2-isopropenyl-5-methyl-4-hexen-1-ol (lavandulol) presumably by 1,4-hydrogen migration of the initially formed diradical intermediate. In acetone the major photoproduct of lavandulol was the corresponding oxetane. In this study *cis–trans* isomerization was not detected. Benzophenone sensitization (Sasaki *et al.*, 1968) of *trans*-chrysanthemic acid produced the *cis*-oxetane among other products, suggesting that sensitized *cis–trans* isomerization had occurred. Isomerization of *t*-butyl-*trans*-chrysanthemate was obtained by acetophenone sensitized photolysis (Ueda and Matsui, 1971). The *cis/trans* ratio at equilibrium was 36/64 whether the substrate was the *cis*- or *trans*-isomer. Similar yields were obtained with the corresponding acid and amide. Direct irradiation of *t*-chrysanthemum dicarboxylic acid resulted in formation of the *cis*-isomer; the process could be enhanced by addition of isobutyrophenone as sensitizer. The possibility of triplet processes in the dicarboxylic acid is enhanced relative to chrysanthemic acid by extension of the conjugated system, thus increasing absorptivity and competition for radiation with the cyclopentenolone chromophore (Figure 9). The absence of *cis–trans* isomerization process in thin films of

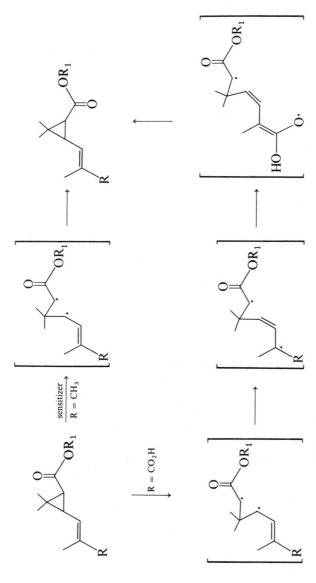

Figure 9 *cis–trans* Isomerization in chrysanthemates and pyrethrates

$hv \longrightarrow$ 1RS-*cis*-E (49%)
1RS-*trans*-E (17%)
1RS-*cis*-Z (2%)
1RS-*trans*-Z (2%)

Figure 10 Isomeric products of kadethrin. Reproduced with permission from Ohsawa and Casida, *J. Agr. Food Chem.*, **27**, 1112–1120, © 1979 American Chemical Society

pyrethrin, allethrin, tetramethrin and dimethrin (Chen and Casida, 1969) and in resmethrin (Ueda *et al.*, 1974) suggests that absorption by the alcohol moiety predominates. However, a recent study showed the presence of *cis*-allethrin in photolysis mixtures arising from *trans-S*-bioallethrin (Ruzo *et al.*, 1980). Separation and identification was achieved utilizing capillary gas chromatography combined with thin-layer chromatography. The formation of *cis*-fenothrin from the *trans*-isomer has been observed (Ruzo, 1980) but in this case the alcohol moiety is a relatively unreactive chromophore.

Kadethrin, which contains a thiolactone ring in conjugation with the vinyl side chain and thus resembles a pyrethrate, undergoes efficient *cis–trans* isomerization of the cyclopropane ring as well as *E–Z* interconversion about the double bond (Ohsawa and Casida, 1979) upon sunlight photolysis (Figure 10). These processes are also apparent from the ester cleavage products identified. The major photoprocesses in kadethrin in the absence of oxygen are isomerizations, as established by using the $[d]$2-octyl ester as a model (Figure 11). These reactions occur via the triplet of the ester carbonyl followed by homolytic cleavage of the C_1–C_3 bond and subsequent recombination. *E–Z* isomerization in the side chain is also presumed to involve excitation of the α,β-unsaturated thiolactone carbonyl to form a diradical intermediate for each *cis*- and *trans*-compound. This allows rotation of the side chain and an *E–Z* isomer mixture is obtained on reforming the original conjugated double bond system. It is also possible that both chromophores are involved via diradical structures (Figure 12). The isomer ratios obtained using either acetone ($E_t = 80$ kcal. mole^{-1}) or isobutyrophenone ($E_t = 73$ kcal. mole^{-1}) are similar. Kadethrin also yields a lactone similar to those reported for other chrysanthemates (Sasaki *et al.*, 1970; Bullivant and Pattenden, 1971) as shown in Figure 13. The newer pyrethroid insecticides which are stabilized by substitution of the isobutenyl moiety by a dihalovinyl group, such as permethrin and decamethrin, undergo *cis–trans* isomerization much more efficiently than the chrysanthemates under direct irradiation by ultraviolet or sunlight. The reaction is also sensitized by energy transfer. Thus, irradiation of *cis*- or *trans*-permethrin (Holmstead *et al.*, 1978a) at $\lambda > 290$ nm leads to an equilibrium mixture of the two isomeric esters. The

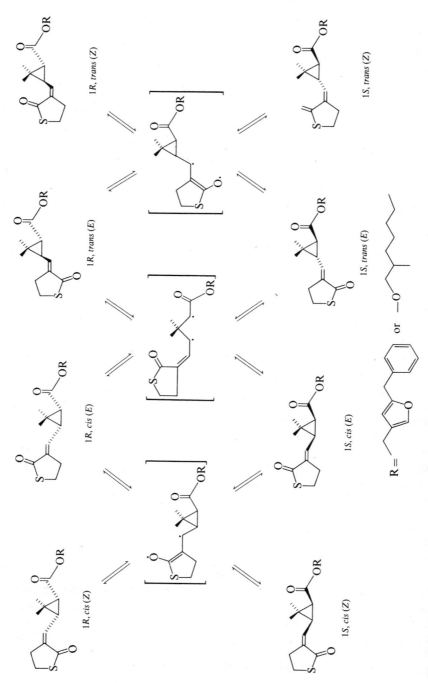

Figure 11 Photoisomerization at the cyclopropanecarboxylate and vinylcarboxylate substituents of kadethrin or its analogous *d*-2-octyl ester. Reproduced with permission from Ohsawa and Casida, *J. Agr. Food Chem.*, **27**, 1112–1120, © 1979 American Chemical Society

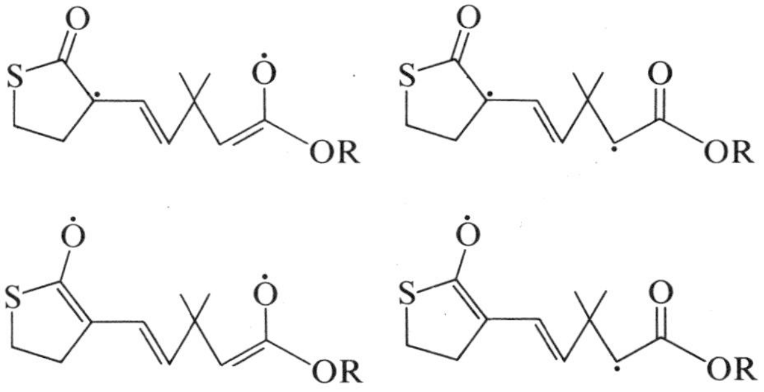

Figure 12 Stabilization of the kadethrin triplet diradical by conjugation leading to isomerized products. Reproduced with permission from Ohsawa and Casida, *J. Agr. Food Chem.*, **27**, 1112–1120, © 1979 American Chemical Society

reaction occurs more rapidly in hexane or aqueous solutions or as a thin film on glass than in methanol or as deposits on silica gel or soil surfaces (Figure 14).

Decamethrin undergoes isomerization to the *trans*-isomer even more efficiently than permethrin (Ruzo *et al.*, 1977); the reaction can be quenched efficiently by piperilene (E_t = 53 kcal. mole^{-1}) or by 1,3-cyclohexadiene (E_t = 50 kcal. mole^{-1}). The *trans*-isomer is the major product at $\lambda > 290$ nm or under sunlight in the solid phase or in a viscous solvent, i.e. 2-propanol. This effect may be the result of radical recombination which minimizes the yield of products from free radical reactions. Carbene photoelimination from permethrin and decamethrin yields dimethyl acrylates in a manner similar to the chrysanthemates (Figure 8). The methyl esters of *cis*-dichloro- and dibromovinyl

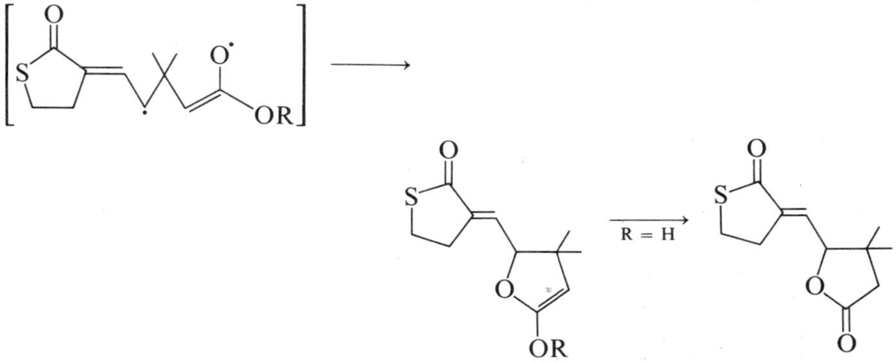

Figure 13 Lactonization of the acid moiety of kadethrin. Reproduced with permission from Ohsawa and Casida, *J. Agr. Food Chem.*, **27**, 1112–1120, © 1979 American Chemical Society

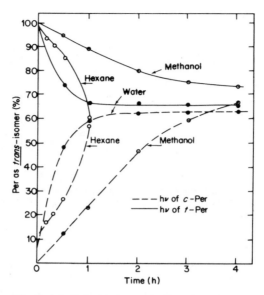

Figure 14 Relative isomerization rates leading to equilibrium *cis–trans* mixtures of (1RS)-*cis*- or (1RS)-*trans*-permethrin in methanol (○), hexane (○) and water (●). Reproduced with permission from Holmstead *et al.*, *J. Agr. Food Chem.*, **26**, 590–595, © 1978 American Chemical Society

cyclopropanecarboxylic acid (Figure 15) yield the corresponding *trans*-isomers when photolysed at 250 and 300 nm (Ruzo and Casida, 1980). The quantum yield of the dibromo-analogue is nearly 20-fold that of the dichlorovinyl methyl ester (Table 1). The reaction rate of the latter compound can be increased several fold by irradiating in the presence of increasing concentrations of n-propyl bromide in methanol. Heavy-atom substitution (i.e. Br) facilitates inter-system crossing affecting excited state lifetimes and phosphorescence quantum yields (Miller *et al.*, 1977). This heavy-atom effect may explain the increased reaction rate and *cis–trans* isomerization in decamethrin relative to the chloro-substituted pyrethroids. Approximate values of the triplet energies of the dibromo- and dichloro-substituted methyl esters have been determined by use of energy transfer agents. The dibromo-esters are quenched below 55 kcal. mole^{-1} while their chlorinated analogues have triplets that lie above 60 kcal. mole^{-1}, i.e. naphthalene ($E_t = 60$ kcal. mole^{-1}) sensitizes the decamethrin acid moiety but quenches that of cypermethrin (Ruzo and Casida, 1980).

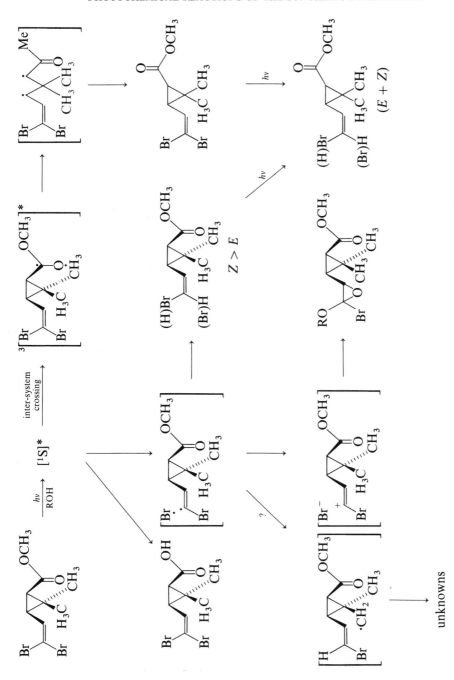

Figure 15 Reactions of methyl *cis*-2,2-dimethyl-3-(2,2-dibromovinyl)-cyclopropanecarboxylate in methanol at 300 nm. Similar processes occur in its chloro-analogue at lower wavelengths. Reproduced from Ruzo and Casida, 1980, with permission of the Royal Society of Chemistry

Isomerization in the alcohol moiety

The substituted cyclopentenone chromophore in allethrin undergoes photo-chemical rearrangement ($\lambda < 300$ nm) in hexane solution to the cyclopropyl derivative (Figure 16) (Bullivant and Pattenden, 1976; Ruzo et al., 1980). The chrysanthemate moiety is not involved in this reaction since allethrolone acetate undergoes the same process. The mechanism of conversion involves a di-π-methane rearrangement, sensitization and quenching experiments suggest that the resulting vinylcyclopropane compound is formed via a triplet excited state. Although this is the major process at low wavelengths it has also been observed under sunlight irradiation in smaller yield (Ruzo et al., 1980). The pyrethrin jasmolin I (Figure 17) does not undergo the di-π-methane rearrangement upon irradiation but yields instead the E-trans-isomer from the Z-cis-2-pentenyl group (Bullivant and Pattenden, 1976). Neither allethrin nor jasmolin I have been found to yield lactones or dimethyl acrylates upon irradiation.

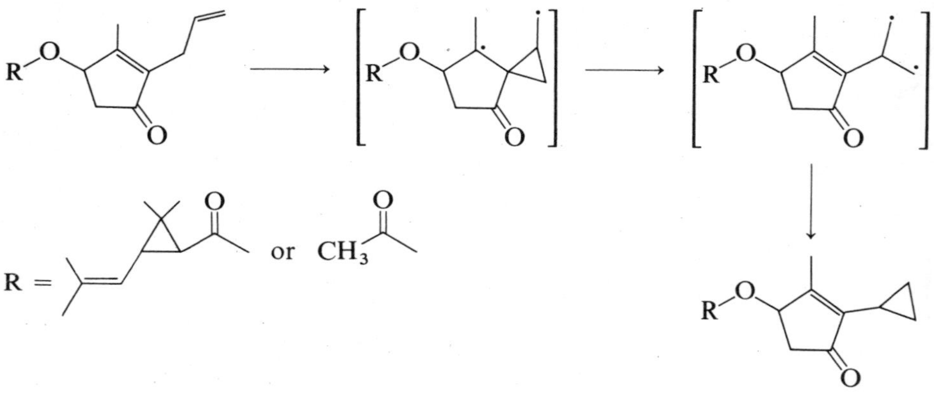

Figure 16 Di-π-methane rearrangement in allethrin and allethrolone acetate solutions

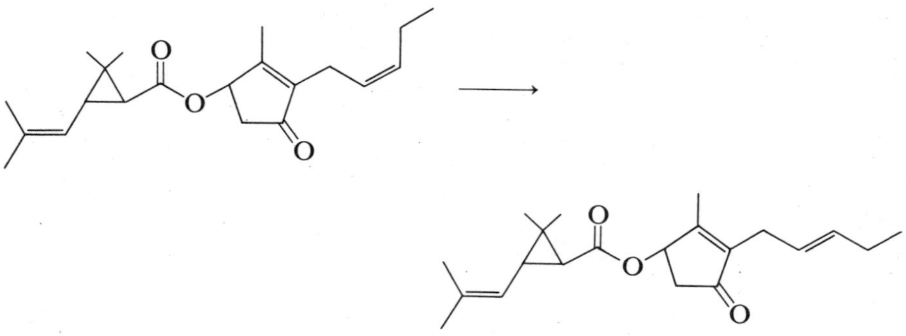

Figure 17 Isomerization of jasmolin I in hexane solution

Reductive dehalogenation

Many halogenated pesticides undergo dehalogenation reactions readily at low wavelengths (Zabik *et al.*, 1976) but not under sunlight irradiation because of their low absorptivity at $\lambda > 290$ nm. The dihalovinyl side chain in the newer pyrethroids is a special case since the double bond is probably conjugated with the ester carbonyl via the bent cyclopropane bonds which resemble an sp^2 system. Thus, *cis–trans* isomerization and reductive dehalogenation are competitive processes in the acid moiety of decamethrin (Ruzo *et al.*, 1977, 1980) at 254 and 300 nm. Cleavage of the carbon–halogen bond does not appear to proceed from a triplet state since the reaction cannot be quenched. Such processes are usually the result of bond homolysis which yields a free radical pair in a solvent cage, a further step may involve electron transfer to form halide and vinyl cations (McNeely and Kropp, 1976; Suzuki *et al.*, 1976) (Figure 15). The free radical can then abstract hydrogen from a proton donor (DH), the vinyl cation may react with a nucleophile (N) to give substitution products (Equation 4).

$$\begin{array}{c}
X_2C{=}CHR \longrightarrow [X \cdot\cdot CX{=}CHR] \\
\end{array}$$

$$\begin{array}{ccc}
 & X^{-\,+}CX{=}CHR & \xrightarrow{\ N\ } & NCX{=}CHR \\
[X\cdot\cdot CX{=}CHR] & & & \\
 & X\cdot + \cdot CX{=}CHR & & \\
\end{array}$$

$$\downarrow \text{RH}$$

$$HX + HCX{=}CHR \tag{4}$$

Photolysis of permethrin in methanol, hexane or aqueous solutions yields the monochlorinated derivative of the parent ester and of the acid moiety, but always as minor products (Figure 18). Their yields are somewhat increased at $\lambda < 300$ nm. In decamethrin (Ruzo *et al.*, 1977) debromination to *E* and *Z* isomers is more efficient because of the decreased bond strength of C–Br relative to C–Cl. Only the free radical products are observed in all solvents. However, it is possible that in methanol or water, products containing oxygen at the vinyl carbon are formed by ionic reactions but are not detected due to instability. *Trans*-debromination is preferred by a factor of four, over *cis*-debromination in methanol. This preference has also been detected in the corresponding methyl ester and it appears to be due to intramolecular reactions by the radical formed upon *cis*-debromination thus decreasing the yield of *E*-isomer (Figure 15) (Ruzo and Casida, 1980). The Z/E ratio is greatest in the order methanol > ethanol > 2-propanol, although the over-all reaction rate does not differ greatly. A possible explanation for this is that intermolecular,

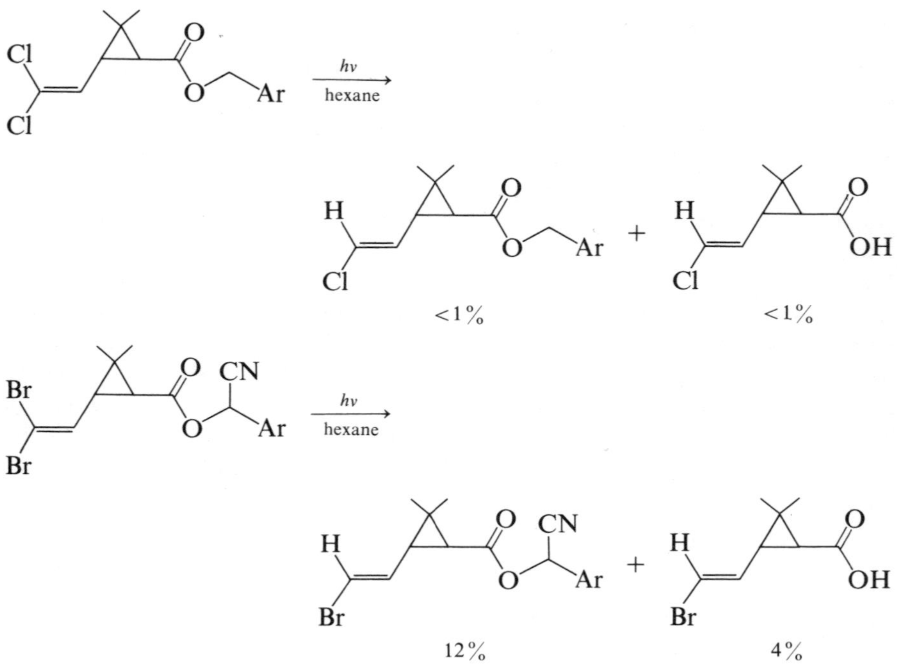

Figure 18   Reductive dehalogenation of permethrin and decamethrin. Reproduced with permission from Ruzo *et al.*, *J. Agr. Food Chem.*, **25**, 1385–1394 © 1977 American Chemical Society and Holmstead *et al.*, *J. Agr. Food Chem.*, **26**, 590–595, © 1978 American Chemical Society

rather than intramolecular abstraction predominates in better hydrogen donors. The ionic mechanism may operate in the formation of the bromoalkoxy epoxide (Figure 15).

Debromination

Bromination of decamethrin and cypermethrin yields the corresponding 1′,2′-dibromoderivatives (tralomethrin and tralocythrin) with insecticidal activities similar to those of the parent materials. Upon photolysis these compounds yield debrominated and dehydrobrominated compounds, *trans*-decamethrin and *trans*-cypermethrin and ester cleavage products similar to those obtained from their precursors (Ruzo and Casida, 1981). The debromination mechanism involves cleavage of a C–Br bond from the most highly halogenated site yielding a radical pair which can either recombine or separate by abstraction of hydrogen from the solvent; collapse of the resulting trihaloethyl radical by extrusion of a second bromine atom yields decamethrin or *cis*-cypermethrin respectively (Figure 19). The yield of debrominated material is

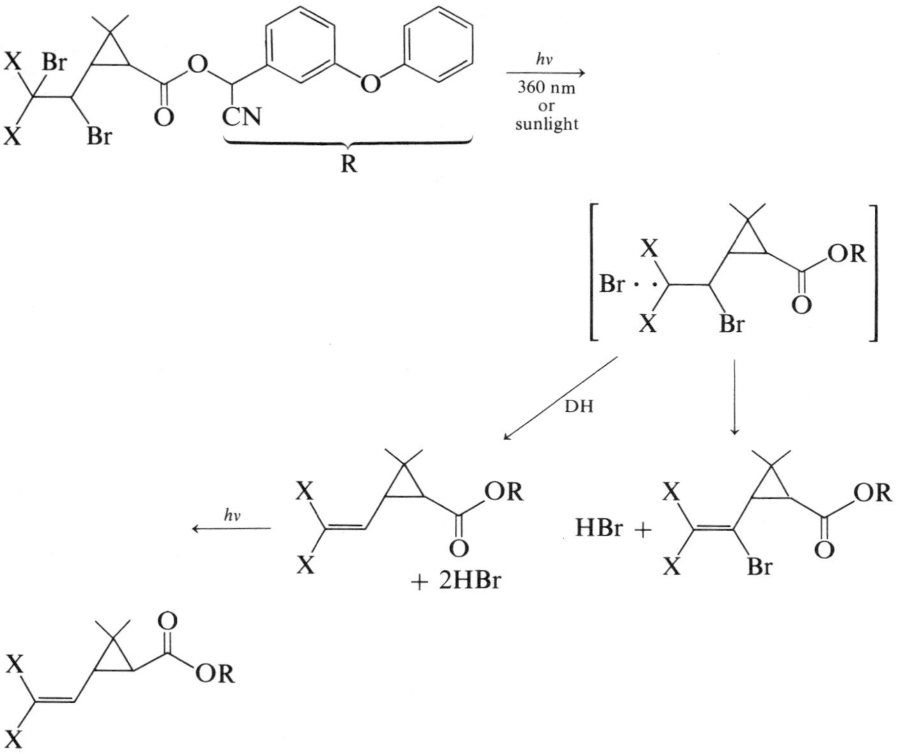

Figure 19   Major photoprocesses of tralomethrin (X = Br) and tralocythrin (X = Cl) in the presence of hydrogen donors (DH)

proportional to the hydrogen-donating ability of the solvent, irradiation in solvents containing no abstractable hydrogens (i.e. benzene) leads to complex mixtures of polar or polymeric materials (Table 2) and decreased yields of debrominated products. Dehydrobromination may occur by intramolecular abstraction of hydrogen at the 1'-position. The *trans*-decamethrin obtained in

Table 2   Ester photoproducts of tralomethrin in solution

Solvent or additive	Percentage reacted	Ester photoproduct yield (%)		
		$-Br_2$	$-HBr$	*trans*-isomer
Methanol	34.3	62.6	3.6	11.7
2-Propanol	32.5	63.6	3.2	12.5
Benzene	42.8	8.6	5.2	1.5
Isopropylbenzene (5% in benzene)	37.9	74.9	4.0	9.5

these studies is believed to arise from light absorption by decamethrin formed during the reaction. The reactions of tralomethrin and tralocythrin in the solid phase upon sunlight irradiation result in product mixtures resembling those obtained in benzene solution.

PHOTO-OXIDATION

Chrysanthemates

Pyrethroids that are chrysanthemate esters are very susceptible to oxidation when photolysed in the presence of oxygen (Chen and Casida, 1969). Several mechanisms operate resulting in oxidized products that resemble those obtained from metabolic systems (Figure 2). A recent study of the photolysis of S-bioallethrin (Ruzo et al., 1980) confirmed these findings and identified additional products arising from epoxidation of the isobutenyl side chain, in analogy with the oxidation of tetramethrin and other chrysanthemates in mammals (Smith and Casida, 1980). When S-bioallethrin is irradiated in solution or in the solid phase (360 nm) in the presence of oxygen, considerable amounts of epoxides from the substrate and its photoproducts are formed (Figure 20). Hydroxylation of the t-methyl group on the isobutenyl moiety also takes place. Formation of the epoxides upon sunlight irradiation is greater in benzene (33%) than in hexane (12%) or aqueous solutions (8%), they are not detected in methanol. The reactions of S-bioallethrin can be sensitized by energy transfer from benzo-phenone, thus increasing the yields of triplet products, i.e. cis-isomer and vinyl cyclopropyl from di-π-methane rearrangement, but the effect of benzophenone

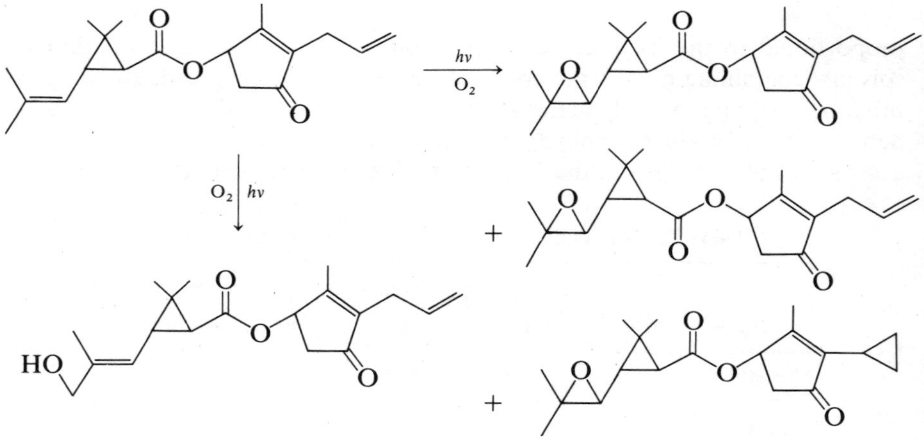

Figure 20   Photo-oxidation of S-bioallethrin. Reproduced with permission from Ruzo et al., J. Agr. Food Chem., **28**, 246–249, © 1980 American Chemical Society

Table 3 Effect of oxygen, sensitizers and a quencher on the photoreactions of S-bioallethrin in benzene (360 nm) (Reproduced with permission from Ruzo et al., J. Agr. Food Chem., **28**, 246–249, © 1980, American Chemical Society)

Additive	Yield, % of total product formation	
	Epoxides	Allyl oxidation
None	42.7	9.3
Benzophenone	31.0	16.3
Benzil	54.9	10.5
1,3-Cyclohexadiene	7.8	< 1.0

and especially of benzil, a diketone, goes beyond their role as sensitizers. These ketones also provide a source of radicals in solution (Figure 21) which can abstract hydrogen from the isobutenyl methyl groups to yield the allylic alcohol, or transfer oxygen to form epoxides. Benzil is considerably more efficient in the latter process (Table 3) (Shimizu and Bartlett, 1976). Epoxidation and allylic oxidations also take place in the absence of sensitizers presumably via radical reactions with triplet oxygen, since these processes are not detected in methanol or 1,3-cyclohexadiene solutions where readily abstractable hydrogen is available to terminate radicals, or in the presence of singlet oxygen. Kadethrin (Ohsawa and Casida, 1979) and resmethrin (Ueda et al., 1974) also undergo epoxidation

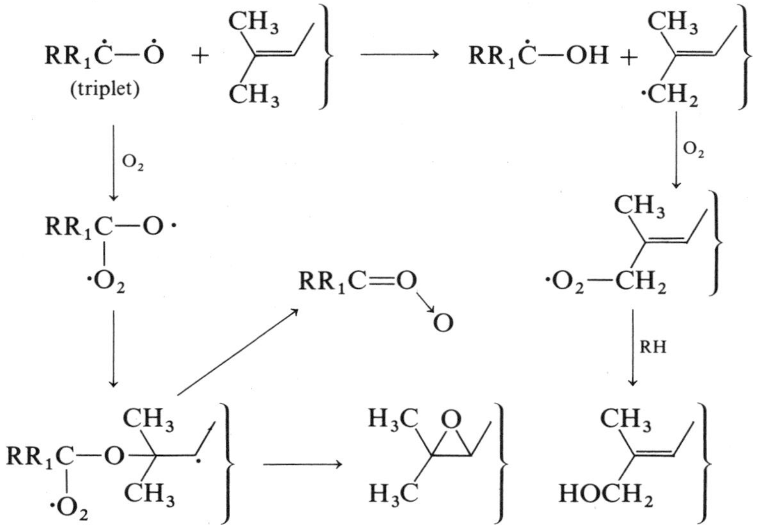

Figure 21 Photoreactions of benzophenone or benzil ($RR_1C{=}O$) with the isobutenyl substituents of S-bioallethrin. Reproduced with permission from Ruzo et al., J. Agr. Food Chem., **28**, 246–249, © 1980 American Chemical Society

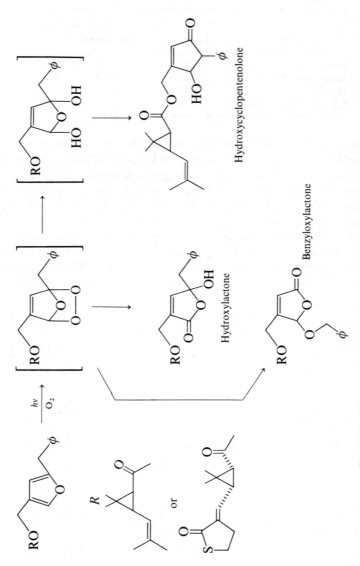

Figure 22 Photo-oxidation of the kadethrin and resmethrin alcohol moiety

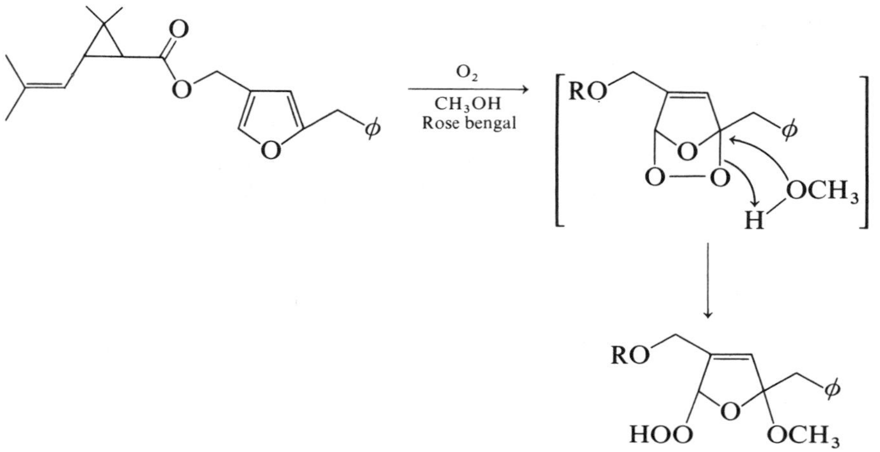

Figure 23 Addition of singlet oxygen to the furyl group of resmethrin. Reproduced with permission from Ueda et al., J. Agr. Food Chem., **22**, 212–220, © 1974 American Chemical Society

in the acid moiety. However, the major oxidative reaction in these pyrethroids involves degradation of the furan ring to yield a cyclic ozonide peroxide by addition of oxygen across the unsaturated system (Figure 22) similar to others postulated as intermediates in the photochemistry of furans (Foote et al., 1967). Migration of the benzyl cation or radical gives the benzyloxylactone; migration of a proton or hydrogen atom from the position symmetrical to the benzyl group gives the hydroxylactone. The hydroxy-cyclopentenolone, obtained as a major product only from resmethrin, may be formed by reduction of the cyclic peroxide to the diol, followed by rearrangement. It is not detected with kadethrin.

The photo-oxidation of the furan ring is probably mediated by singlet oxygen (Ueda et al., 1974; Foote et al., 1967) since rose bengal sensitization in methanol yields a methoxy-hydroperoxide (Figure 23) by reaction of the solvent with the endoperoxide.

Dihalovinylcyclopropanecarboxylates

Permethrin and monochloro-permethrin do not undergo photo-oxidation within seven days of irradiation in methanol containing singlet oxygen. Permethrin does not react with excess m-chloroperbenzoic acid or trifluoro-peroxyacetic acid in dichloromethane at 25 °C (Holmstead et al., 1978a). However, the epoxide of monochloropermethrin is obtained with an excess of oxidant. Decamethrin is also resistant to epoxidation (Ruzo et al., 1977) but its methyl ester yields oxidation products lacking one bromine substituent (Ruzo and Casida, 1980) (Figure 15). Oxidation of the dihalovinyl moiety may be possible under different photolysis conditions since other deactivated alkenes, such as tri- and tetrachloroethylene are epoxidized by irradiation at high temperatures (Kline *et al.*, 1978).

PHOTODECOMPOSITION OF PYRETHROIDS ON LEAF AND SOIL SURFACES

Some of the reactions undergone by pyrethroids in biological systems have photochemical counterparts, i.e. oxidation and ester hydrolysis. Other processes can be clearly assigned to specific mechanisms as are conjugation (metabolic) and *cis–trans* isomerization (photochemical). The pyrethroids are readily degraded in plants with lifetimes in the order of days for the less stable compounds and longer for permethrin (Gaughan and Casida, 1978) and decamethrin (Ruzo and Casida, 1979).

trans- And *cis*-permethrin are degraded on cotton and bean leaves mainly by ester cleavage and oxidation reactions followed by conjugation of the liberated acid and alcohol moieties as glycosides (Gaughan and Casida, 1978; Ohkawa *et al.*, 1977). Some *cis–trans* isomerization at the cyclopropane ring takes place photochemically. Permethrin is slowly degraded on soil, exposure for 48 days in the dark resulting in 35 % loss, while loss in the light is 55 % (Holmstead *et al.*, 1978a). No difference was observed in the amount of dichlorovinylcyclopropanecarboxylate produced upon exposure to sunlight suggesting that ester cleavage is mainly chemical. However, several products observed can be assumed to be formed photochemically (Table 4); trace amounts of reductive dechlorination and carbene photoelimination products are obtained.

The only product obtained from decamethrin on cotton leaves that can be unequivocally assigned to a photochemical process is the *trans*-isomer (Ruzo and Casida, 1979). The yield of this compound increases up to seven days and

Table 4 Degradation products of (1*RS*)-*trans*- and (1*RS*)-*cis*-permethrin following 48 days exposure in the dark or to sunlight on a thin layer of soil (Reproduced with permission from Holmstead *et al.*, *J. Agr. Food Chem.* **26**, 590–595, © 1978 American Chemical Society)

| | [^{14}C]Permethrin equivalents, % | | | |
| | Dark | | Sunlight | |
Compound or product	trans	cis	trans	cis
Permethrin and products retaining ester linkage				
Permethrin				
trans-	49.9	0.4	14.2	0.9
cis-	0.3	49.2	0.4	15.8
Monochloro-permethrin	0.0	0.0	0.1	0.2
3-Phenoxybenzyl dimethylacrylate	0.0	0.0	0.2	0.1

Table 5 Photoisomerization of decamethrin
and cis-permethrin on cotton leaves

	cis/trans Ratio (seven days)	
Pyrethroid	Greenhouse	Field
Decamethrin	5.7	1.4
cis-Permethrin	8.1	3.8

then decreases. However, the *cis–trans* isomerization process in decamethrin is more significant than in *cis*-permethrin (Table 5) presumably due to the heavy-atom effect.

Fenvalerate is degraded on cotton at a rate intermediate between that of decamethrin (least stable) and permethrin (Holmstead *et al.*, 1978b). The major product is decarboxylated fenvalerate.

TOXICITY OF PYRETHROID PHOTOPRODUCTS

The insecticidal activities of the pyrethrins are substantially decreased upon exposure to light (Stahl, 1960; Chen and Casida, 1969; Miskus and Andrews, 1972). Photoproduct mixtures from allethrin (Bullivant and Pattenden, 1971; Chen and Casida, 1969), tetramethrin and dimethrin (Chen and Casida, 1969), resmethrin (Ueda *et al.*, 1974), decamethrin (Ruzo *et al.*, 1977), fenvalerate (Holmstead *et al.*, 1978b) and kadethrin (Ohsawa and Casida, 1979) also exhibit

Table 6 Composition and intraperitoneal toxicity to mice of photodecomposed (+)-*trans*-resmethrin. (Reproduced with permission from Ueda *et al.*, *J. Agr. Food Chem.*, **22**, 212–220, © 1974 American Chemical Society)

Irradiation time, h*	Composition, %		LD_{50}, mg. kg^{-1}
	Resmethrin	(+)-*trans*-chrysanthemic acid	
Photodecomposed resmethrin			
2	78	5	1800
3	39	18	1200
5	43	12	960
15	0	6	580
Comparison compounds			
0	100	0	>2500
0	0	100	98

* The exposure times of 2 h and 5 h were on a hazy day while those of 3 h and 15 h were on clear days

decreased insecticidal potency. Photodecomposition of tralomethrin and tralocythrin does not initially result in a less insecticidal material since the primary photoproducts are themselves potent insecticides (Ruzo and Casida, 1981). In only a few cases have individual photoproducts been assayed for insecticidal activity, thus the dimethyl acrylate and cyclopropyl-derivative of allethrin are not insecticides (Bullivant and Pattenden, 1971). The epoxides of resmethrin and ethano-resmethrin are less toxic to houseflies than the parent pyrethroids (Ueda et al., 1974).

The mammalian toxicity of pyrethroid degradation products is of greater interest from the viewpoint of environmental safety. Again, in most cases only mixtures have been assayed and found to be less acutely toxic than the starting material. However, resmethrin photolysis gives 1R,trans-chrysanthemic acid which is more toxic than resmethrin itself (Ueda et al., 1974) as well as an unidentified toxicant which, in contrast to trans-chrysanthemic acid, builds up with time in the reaction mixture (Table 6). Fenvalerate (Holmstead et al., 1978b) and decamethrin (Ruzo et al., 1977) yield two products that are more acutely toxic to mice than the pyrethroids (Table 7), 3-phenoxybenzyl cyanide and 3-phenoxybenzoyl cyanide; the latter compound probably releases cyanide in vivo upon hydrolysis. The toxicity of these compounds could nŏt be synergized. The long-term mammalian toxicity of pyrethroid photoproducts has not been reported although there are indications of mutagenic activity in allethrin photoproduct mixtures (Casida, 1980; E. C. Kimmel, J. E. Casida and L. O. Ruzo, unpublished results, 1981). A variety of tests on permethrin do not indicate mutagenic activity (Ruzo and Casida, 1977) but individual photoproducts from this and other pyrethroids have not as yet been examined in this respect.

Table 7   Mouse intraperitoneal LD_{50} values for various pyrethroids and their metabolites or degradation products. The abbreviation CA designates chrysanthemic acid and its derivatives

Pyrethroid		Metabolites or degradation products	
Compound	mg. kg^{-1}	Compound	mg. kg^{-1}
[1RS,trans]Permethrin	>1000	[1R,trans]CA	98
[1RS,cis]Permethrin	>1000	[1R,cis]CA	600
[1R,cis]Permethrin	1000	[1R,trans]dichloroCA	210
[1RS,trans]Cypermethrin	>500	[1R,cis]dichloroCA	370
[1RS,cis]Cypermethrin	28	[1R,cis]dibromoCA	525
Decamethrin	10	α-isopropyl-4-chlorophenylacetic acid	>500
[(±)αRS]Fenvalerate		phenoxybenzyl alcohol	575
		phenoxybenzaldehyde	>500
		phenoxybenzoic acid	350
		phenoxybenzoyl cyanide	22
		KCN	6

CONCLUSIONS

Some trends can be deduced from the photochemical studies on pyrethroids to date. It is now understood that because of the seemingly endless structural modifications that retain biological activity the pyrethroids cannot be studied as a class of compounds in regard to their photochemistry. As long as some moieties are present, i.e. the cyclopropane ring, certain photoreactions, such as *cis–trans* isomerization are possible, however, substitution of this group by others can result in completely different degradation reactions, such as decarboxylation in fenvalerate. It is evident that developments resulting in the new classes of pyrethroids have introduced sufficient photostability while preserving high insecticidal potency. Further research is necessary to determine the toxicological characteristics as well as environmental persistence of pyrethroid photoproducts.

ACKNOWLEDGEMENT

The studies carried out at Berkeley were supported in part by the National Institute of Environmental Health Science (PO1ES00049).

REFERENCES

Abe, Y., Tsuda, K., and Fujita, Y. (1972) 'Photostability of pyrethroidal compounds', *Botyu-Kagaku*, **37**, 102–111.

Brooke, J. P. (1967) 'Effect of dioxyphenyl synergists upon stability of the pyrethroids', *Pyrethrum Post*, **9**, 18–30.

Brown, N. C., Hollinshead, D. T., Phipers, R. F., and Wood, M. C. (1957) 'Analysis of pyrethrins', *Soap Chem. Specialities*, **33**, 87–109.

Bullivant, M. J., and Pattenden, G. (1971) 'Photochemical decomposition of chrysanthemic acid and its alkyl esters', *Pyrethrum Post*, **11**, 72–76.

Bullivant, M. J., and Pattenden, G. (1976) 'Photochemistry of 2-(prop-2-enyl)cyclopent-2-enones', *J. Chem. Soc. Perkin I*, 249–256.

Burt, P. E., Elliott, M., Farnham, A. W., Janes, N. F., Needham, P. H., and Pulman, D. A. (1974) 'Geometrical and optical isomers of (2,2-dichlorovinyl)-cyclopropanecarboxylic acid and insecticidal esters with 5-benzyl-3-furyl methyl and 3-phenoxybenzyl alcohols', *Pestic. Sci.*, **5**, 791–799.

Campbell, A., and Mitchell, W. (1950) 'An examination of polymerized pyrethrins', *J. Sci. Food Agr.*, **1**, 137–139.

Casida, J. E. (1980) 'Development of synthetic insecticides from natural products: pyrethroids from pyrethrins', *ICIPE Conference on the Use of Naturally Occurring Plant Products in Pest and Disease Control*, Nairobi, Kenya.

Casida, J. E., Gaughan, L. C., and Ruzo, L. O. (1979) 'Comparative metabolism of pyrethroids derived from 3-phenoxybenzyl and α-cyano-3-phenoxybenzyl alcohols', in *Advances in Pesticide Science*, (H. Geissbühler, Ed.), Pergamon Press, Oxford, Part 2, p. 80.

Chen, Y.-L., and Casida, J. E. (1969) 'Photodecomposition of pyrethrin I, allethrin, phthalthrin and dimethrin', *J. Agr. Food Chem.*, **17**, 208–215.

32   PROGRESS IN PESTICIDE BIOCHEMISTRY

Crowe, T. J., Glynne Jones, G. D., and Williamson, R. (1961) 'Use of pyrethrum formulations to control *Antestiopsis* on coffee in E. Africa', *Bull. Entomol. Res.*, **52**, 31–41.

Elliott, M. (1977) 'Synthetic pyrethroids', *ACS Symp. Ser.*, **42**, 1–28.

Elliott, M., Farnham, A. W., Janes, N. F., Needham, P. H., Pulman, D. A., and Stevenson, J. H. (1973) 'A photostable pyrethroid', *Nature*, **246**, 169–170.

Elliott, M., Farnham, A. W., Janes, N. F., Needham, P. H., and Pulman, D. A. (1974a) 'Synthetic insecticide with a new order of activity', *Nature*, **248**, 710–711.

Elliott, M., Farnham, A. W., Janes, N. F., Needham, P. H., and Pulman, D. A. (1974b) 'Insecticidally active conformations of pyrethroids', *ACS Symp. Ser.*, **2**, 80–91.

Foote, C. S., Wuesthoff, M. T., Wexler, S., Burstain, I. G., Denny, R., Schenck, G. O., and Schulte-Elte, K.-H. (1967) 'Photosensitized oxygenation of alkyl-substituted furans', *Tetrahedron*, **23**, 2583–2599.

Freeman, S. K. (1956) 'Stability of allethrin versus pyrethrins', *Soap Chem. Specialities*, **32**, 131–138.

Gaughan, L. C., and Casida, J. E. (1978) 'Degradation of *trans*- and *cis*-permethrin on cotton and bean plants', *J. Agr. Food Chem.*, **26**, 525–528.

Givens, R. S., and Oettle, W. F. (1972) 'Photodecarboxylation of benzyl esters', *J. Org. Chem.*, **37**, 4325–4334.

Glynne Jones, G. D. (1960) 'Studies on the photolysis of pyrethrum', *Ann. Appl. Biol.*, **48**, 352–368.

Head, S. W., Sylvester, N. K., and Challinor, S. K. (1968) 'Effect of piperonyl butoxide on the stability of pyrethrum films', *Pyrethrum Post*, **9**, 14–22.

Holmstead, R. L., Casida, J. E., and Ruzo, L. O. (1976) 'Photochemical reaction of pyrethroid insecticides', *ACS Symp. Ser.*, **42**, 137–146.

Holmstead, R. L., and Fullmer, D. G. (1977) 'Photodecarboxylation of cyanohydrin esters: models for pyrethroid photodecomposition', *J. Agr. Food Chem.*, **25**, 56–58.

Holmstead, R. L., Casida, J. E., Ruzo, L. O., and Fullmer, D. G. (1978a) 'Pyrethroid photodecomposition: permethrin', *J. Agr. Food Chem.*, **26**, 590–595.

Holmstead, R. L., Fullmer, D. G., and Ruzo, L. O. (1978b) 'Pyrethroid photodecomposition: pydrin', *J. Agr. Food Chem.*, **26**, 954–959.

Kline, S. A., Solomon, J. J., and van Duuren, B. L. (1978) 'Synthesis and reactions of chloroalkene epoxides', *J. Org. Chem.*, **43**, 3596–3600.

Matuszewski, B., Givens, R. S., and Neywick, C. (1973) 'Photodecarboxylation of esters: photolysis of α- and β-naphthyl derivatives, *J. Amer. Chem. Soc.*, **95**, 595–596.

McNeely, S. A., and Kropp, P. J. (1976) 'Photochemistry of alkyl halides. Generation of vinyl cations', *J. Amer. Chem. Soc.*, **98**, 4319–4320.

Miller, J. C., Meek, J. S., and Stickler, S. J. (1977) 'Heavy atom effects on the triplet lifetimes of naphthalene and phenanthrene', *J. Amer. Chem. Soc.*, **99**, 8175–8179.

Minamite, Y., Hirobe, H., Ohgami, H., and Katsuda, Y. (1978) 'A new pyrethroid: alkoxycyclopropanecarboxylic acid ester, insecticidal activity and photostability', *J. Pestic. Sci.*, **3**, 437–439.

Miskus, R. P., and Andrews, T. L. (1972) 'Stabilization of pyrethrins and allethrin', *J. Agr. Food Chem.*, **20**, 313–315.

Ohkawa, H., Kaneko, H., and Miyamoto, J. (1977) 'Metabolism of permethrin in bean plants', *J. Pestic. Sci.*, **2**, 67–76.

Ohno, N., Fujimoto, K., Okuno, Y., Mizutani, T., Hirano, M., Itaya, N., Honda, T., and Yoshioka, H. (1974) 'A new group of synthetic pyrethroids not containing cyclopropanecarboxylates', *Third International Congress of Pesticide Chemistry (IUPAC)*, Helsinki, Finland, Abstract p. 346.

Ohsawa, K., and Casida, J. E. (1979) 'Photochemistry of the potent knockdown pyrethroid kadethrin', *J. Agr. Food Chem.*, **27**, 1112–1120.

Ruzo, L. O. (1980) 'Photodecomposition and toxicology of pyrethroids', *Roussel-Uclaf Table Ronde on Pyrethroid Chemistry and Toxicology, Paris*, **37**, 42.

Ruzo, L. O., and Casida, J. E. (1977) 'Metabolism and toxicology of pyrethroids with dihalovinyl substituents', *Environ. Health Perspec.*, **21**, 285–292.

Ruzo, L. O., and Casida, J. E. (1979) 'Degradation of decamethrin on cotton plants', *J. Agr. Food Chem.*, **27**, 572–575.

Ruzo, L. O., and Casida, J. E. (1980) 'Mechanistic aspects in reaction of the dihalogeno-vinyl cyclopropanecarboxylate substituent', *J. Chem. Soc. Perkin I*, 728–732.

Ruzo, L. O., and Casida, J. E. (1981) 'Pyrethroid photochemistry: (*S*)-α-cyano-3-phenoxy-benzyl *cis*-(1*R*)-3-(2',2',2'-bromodihalo-1'-bromoethyl)-2,2-dimethylcyclopropanecar-boxylates', *J. Agr. Food Chem.*, **29**, 702–706.

Ruzo, L. O., Holmstead, R. L., and Casida, J. E. (1976) 'Solution photochemistry of the potent pyrethroid insecticide decamethrin', *Tetrahedron Letters*, **1976**, 3045–3048.

Ruzo, L. O., Holmstead, R. L., and Casida, J. E. (1977) 'Pyrethroid photochemistry: Decamethrin', *J. Agr. Food Chem.*, **25**, 1385–1394.

Ruzo, L. O., Gaughan, L. C., and Casida, J. E. (1980) 'Pyrethroid photochemistry: S-Bioallethrin', *J. Agr. Food Chem.*, **28**, 246–249.

Sasaki, T., Eguchi, S., and Ohno, M. (1968) 'Some reactions of the isobutenyl group in chrysanthemic acid', *J. Org. Chem.*, **33**, 676–679.

Sasaki, T., Eguchi, S., and Ohno, M. (1970) 'Photochemical behavior of chrysanthemic acid and its derivatives', *J. Org. Chem.*, **35**, 790–793.

Shimizu, N., and Bartlett, P. D. (1976) 'Photo-oxidation of olefins sensitized by α-dike-tones and benzophenone. A practical epoxidation method with biacetyl', *J. Amer. Chem. Soc.*, **98**, 4193–4200.

Smith, I., and Casida, J. E. (1980) 'Formation and degradation of epoxychrysanthemates', *Tetrahedron Letters*, **22**, 203–206.

Stahl, E. (1960) 'Zur Inaktivierung der Pyrethrine am Wirkungsort', *Arch. Pharm.*, **293/65**, 531–537.

Suzuki, T., Sonoda, T., Kobayashi, S., and Taniguchi, H. (1976) 'Photochemistry of vinyl bromides: a novel 1,2-aryl migration', *J.C.S. Chem. Comm.*, **1976**, 180.

Ueda, K., and Matsui, M. (1971) 'Photochemical isomerization of chrysanthemic acid and its derivatives', *Tetrahedron*, **27**, 2771–2774.

Ueda, K., Gaughan, L. C., and Casida, J. E. (1974) 'Photodecomposition of resmethrin and related pyrethroids', *J. Agr. Food Chem.*, **22**, 212–220.

Van Dusen, W., and Hamill, W. H. (1962) 'Ionic and free radical processes in the radiolysis and sensitized photolysis of benzene solutions', *J. Amer. Chem. Soc.*, **84**, 3648–3658.

Yamamoto, I., and Casida, J. E. (1966) '*O*-Demethyl pyrethrin II analogs from oxidation of pyrethrin I, allethrin, dimethrin and phthalthrin by a house fly enzyme system', *J. Econ. Entomol.*, **59**, 1542–1543.

Yamamoto, I., Kimmel, E. C., and Casida, J. E. (1969) 'Oxidative metabolism of pyre-throids in houseflies', *J. Agr. Food Chem.*, **17**, 1227–1236.

Zabik, M. J., Leavitt, R. A., and Su, G. C. C. (1976) 'Photochemistry of bioactive com-pounds: a review of pesticide photochemistry', *Ann. Rev. Entomol.*, **21**, 61–79.

Zimmerman, H. E., and Sandel, V. R. (1963) 'Mechanistic organic chemistry II. Solvolytic photochemical reactions', *J. Amer. Chem. Soc.*, **85**, 915–922.

Progress in Pesticide Biochemistry, Volume 2
Edited by D. H. Hutson and T. R. Roberts
© 1982 John Wiley & Sons Ltd.

CHAPTER 2

The potential of stable isotopes in pesticide metabolism studies

P. Hendley

INTRODUCTION

The last decade has seen a dramatic increase in the application of stable isotopes to biological problems (Klein and Klein, 1978 and 1979). This has largely been due to improvements in the available technology for stable isotope detection and quantification allied with the increasing availability of suitably labelled substrates. Because of the ethical advantages of avoiding the use of radiochemicals in humans, the pharmaceutical chemists have been the leaders in the use of stable isotopes in metabolism studies (Klein et al., 1978). In the pesticide field, where only limited metabolism studies in man are carried out, the cost/benefit equation for the use of stable isotopes must be carefully assessed. In the near future it is most unlikely that the detection and quantification of unknown metabolites of stable labelled compounds will become simpler, quicker or cheaper than the techniques currently available for measuring radioisotopes. Why then should the pesticide chemist even consider stable labels? What can the technique offer that radiolabelling does not provide? There are five aspects which I will summarize here and return to later.

1. Stable isotopes offer the potential of obtaining considerably more structural information on metabolites. When stable labelled compounds are mixed with unlabelled compound, a mass spectral tag in the form of a characteristic ion cluster is generated; this applies to the parent pesticide and to any metabolites and mass spectral fragments bearing the isotopic atoms. Hence ions due to metabolites and their fragments may be easily distinguished from general background in mass spectra, a task that often complicates the interpretation of the spectra of many radiolabelled samples.
2. Stable labelling with ^{13}C or ^{15}N gives the option of using nuclear magnetic resonance (n.m.r.) which has the advantage of being non-destructive. The spectra can give useful information on chemical changes occurring at or near the enriched atoms during metabolism. Additionally the physical environment of the labelled atom can be probed to yield information on enzyme sites or bound residues.
3. The availability of a stable labelled compound offers the option of simplified residue analysis via its use as an internal standard or a residue carrier. Similarly the various normal and reverse isotope dilution techniques can be invaluable in metabolism studies.
4. Stable isotopes offer the potential of an environmentally acceptable alternative to radiochemicals when necessary, e.g. in large scale field trials or certain large animal studies.
5. Stable isotopes also offer an additional tool to help tackle some of our more complex problems, e.g. stereoselective metabolism, formulation efficacy, bound residues etc. Experience shows that the chances of success improve as the number of approaches available increases.

This chapter will draw together some of the available information in these areas to help pesticide chemists judge whether savings, in terms of time or of improved results, could be achieved for their problems by using stable isotopes. In addition, two excellent reviews centred around drug metabolism should be consulted (Hawkins, 1977; Draffan, 1978). There are inevitably some applications which are beyond the scope of this review. Paramount among these is the application of ^{15}N for agronomic and agricultural studies which has been fully covered in reviews by Hauck and Bystrom (1970, 1981). Similarly the increasingly vast literature available on stable isotopes as internal standards for quantification can only be briefly summarized to illustrate relevant areas. The stable isotope reference compilations of Klein and Klein (1978, 1979) give more comprehensive access to this data.

GENERAL BACKGROUND INFORMATION AND EXPERIMENTAL DESIGN

Isotopes

Isotopes may be simply defined as forms of the same element which differ purely in the numbers of neutrons per nucleus. The numbers of protons and electrons are, by definition, fixed and so the stability of the nuclei depends on their ability to lose energy by emitting charged particles. Table 1 lists the common isotopes of interest to pesticide chemists and records their percentage abundance relative to the normal isotope—henceforth the 'normatope'.

The existence of stable isotopes was established in the late 1920s. They rapidly became commercially available and until the late 1940s were used quite extensively in biochemistry. However, coincident with the Manhattan project, radio-labelled tracers become feasible and since the detection and quantification of radiolabels was easier (and rapidly became even more so) the use of ^2H and ^{13}C in particular among the stable isotopes waned until the late 1960s. ^{15}N And ^{18}O have, of course, continued to be used where necessary due to the lack of a suitable radioisotope. The ability to 'trace' these elements is a particular advantage of the stable isotope technique.

Table 1 Isotopes of elements of pesticidal interest (Gordon and Ford, 1972)

Normatope	Stable isotope (% relative abundance)	Radioisotope
^1H	$^2H(0.015)$	^3H
^{12}C	$^{13}C(1.11)$	^{14}C
^{14}N	$^{15}N(0.37)$	—
^{16}O	$^{18}O(0.204)$	—
	^{17}O —	—
^{32}S	$^{34}S(4.22)$	^{35}S
^{35}Cl	$^{37}Cl(24.47)$	^{36}Cl
^{79}Br	$^{81}Br(49.46)$	^{82}Br
^{120}Sn	9 isotopes	8 isotopes

Isotope effects

The essential assumption upon which all isotopic labelling is based is that the labelled molecule will behave in the same way as normatopic material, i.e. that there will be no isotope effect. Unfortunately this assumption is not always justified; where bond formation or cleavage is involved, the difference in masses and binding energies between isotopes can lead to differences in reaction rates. For the heavier elements, the magnitude of these effects is generally small enough to make little difference to metabolic or photochemical processes and thus the assumption is valid in practice. For the isotopes of hydrogen, however, there is a mass difference of 100 % or 200 % and, additionally, bonds to hydrogen are frequently involved in pesticide chemistry. In some cases, even chromatographic resolution of normatopic and ^{2}H-labelled chemicals have been observed (Horning et al., 1979b). Hence the isotope effects can be considerable and care must be taken in interpreting results. Isotope effects have been turned to the advantage of the pharmacologist; by introducing ^{2}H into drugs, one metabolic pathway can be slowed down and a more useful route can be accentuated. This gives another measure of the potential significance of isotope effects (Horning et al., 1979b; Jarman and Foster, 1979).

It is the marked isotope effect with ^{2}H that leads to its relative toxicity. While rats have been shown to tolerate 5 % of ^{2}H$_2$O in drinking water for up to seven months with no deleterious effects, 20 % concentrations gave noticeable changes (Hawkins, 1977). Peppermint (Mentha piperita) plant cuttings were grown in ^{2}H$_2$O nutrient solutions and a significant growth inhibition effect was noted when the ^{2}H$_2$O level reached 50 %. A 70 % level caused growth to cease altogether and at all ^{2}H$_2$O levels flowering was impaired. Interestingly, the plant growth inhibitor, maleic hydrazide, actually stimulated growth in ^{2}H$_2$O inhibited plants. Similar effects were seen with A. belladonna and in this species the seed germination rate was 0 % with water containing more than 50 % ^{2}H$_2$O (Blake et al., 1975). The fact that there was no reported toxicology with mice fed ^{13}C-enriched diets, even though incorporations of the order of 60 atom % ^{13}C were recorded after 33 weeks, is equally a reflection of the smaller isotope effects noted with the heavier atoms.

Availability of starting materials

The demand for and availability of stable labelled materials have been closely linked. Consequently there are now several sources in Europe and the United States of America offering a wide range of synthetic precursors and readily available stable labelled compounds. In addition the major companies are happy to tender for custom synthesis. Normally the isotopic precursors will have to be further synthesized to the desired compound and therefore, with the exception of ^{2}H, many of the usual problems well known to the ^{14}C radiochemist are encountered. There is, however, a difference of scale. If the isotopic material is at any stage to be used as an internal standard for g.c./m.s. analyses, no dilution with unlabelled compounds is possible. This does not pose the problem it would in radiosynthesis since one works on a weight basis with stable isotopes rather than from the more familiar specific activity standpoint.

Stable labelling routes tend to fall between the radiochemical and normal synthetic scales and this may mean that checking the feasibility of scaling *up* a radiosynthesis is necessary! Generally, however, the methods are compatible and so the radiochemist's time can be used efficiently by incorporating the synthesis of the stable label into the 'cold' method-evaluation work. Sometimes for reasons of economy or unavailability of a suitable precursor, synthesis from one-carbon precursors may be required. This poses some unusual problems but a recent paper (Ott, 1979) describes some useful methods and a constructive approach to synthetic design in this area.

Isotope exchange

^{2}H-Labelling leads to the problem (well known with ^{3}H-labels) of label exchange during use of the material; consequently care must always be taken to guard against this. Good assay technique requires that label stability be confirmed for any procedure and a good example is [17α, 21, 21-^{2}H$_3$-]tetrahydrodeoxy-corticosterone (**1**) which was prepared by ^{2}H-exchange. A sample was run

(1)

through the complete assay procedure along with a sample of an underivatized epimer (different g.c. retention time) as an internal comparison. Gas chromato-graphy/selected ion monitoring was used to compare the responses of the mixture before and after running the sample through the process. The results showed that no exchange of ^{2}H occurred and also that the ratios of the compounds were unaltered by the processing; indicating the absence of an isotope effect (Shackleton *et al.*, 1979).

Purity

One important advantage of stable isotope labelled chemicals is their relative stability. There is no need to consider the possibility of radiolysis that bedevils the storage of high specific activity radiochemicals (Evans, 1976). Nevertheless, the usual criteria of chemical purity must not be ignored when relying on mass spectrometry. To guard against the presence of involatile chemical impurities, a simple test is to compare the g.c./m.s. responses of the sample and an analytically verified standard on a weight for weight basis (Horning et al., 1979a). The isotopic purity of stable isotopically labelled materials is normally expressed as atom-% of the isotope; this normally depends on the source of the isotope. As the demand for stable labels has increased, so has the purity range available to the user. ^{13}C Is produced by cryogenic fractional distillation of carbon monoxide (another application of isotope effects) and isotopic purities of up to 99 atom-% are now commercially available. Both ^{18}O and ^{15}N can now be obtained at up to 99.9 atom-% (e.g. see Lockhart, 1980).

Position of labelling

Synthetic routes, potential precursors and, indeed, the information available from the final experiment are all determined by the choice of labelling position within the pesticide molecule. The same parameters are involved as in radio-labelling experiments viz.

1. What is the position that will give maximum information?
2. Will the label be stable in this position for most of the expected metabolic transformations?
3. Are the alternative labelling positions cheaper and nearly as effective at answering questions 1 and 2?

We have already concluded that it is most unlikely that stable labelling will be used for metabolism studies instead of radiolabelling. ^{14}C-Material will probably already be available and so one must decide whether a ^{13}C-stable label should be introduced at the same position as a ^{14}C-label or elsewhere. (If the same precursor is to be used, the position is the same by definition.) The option of gaining extra information by labelling elsewhere is attractive, although great care is required in interpreting results with mixtures of the two labels.

It is also of particular importance with stable labels to consider whether the proposed labelled atom or atoms will be retained in most of the significant mass spectral fragments. If not, the potential for metabolite identification by re-cognition of series of ion clusters will be significantly reduced. A good example of this is provided in the first metabolism study to employ the 'isotope cluster' technique (Knapp et al., 1972). Nortriptyline (2) was labelled with deuterium as shown (M^+ of derivative, $m/e = 362$) and an equimolar mixture of normatope and isotope was administered to a rat. Collection of urine, extraction and derivatization/g.c./m.s. revealed a metabolite with a characteristic $M/M + 3$ doublet at m/e 357/360 due to compound (3). However, as the label was in a less than ideal position, the doublet was only visible in the molecular ion. A better position would have given a series of doublets throughout the mass spectra.

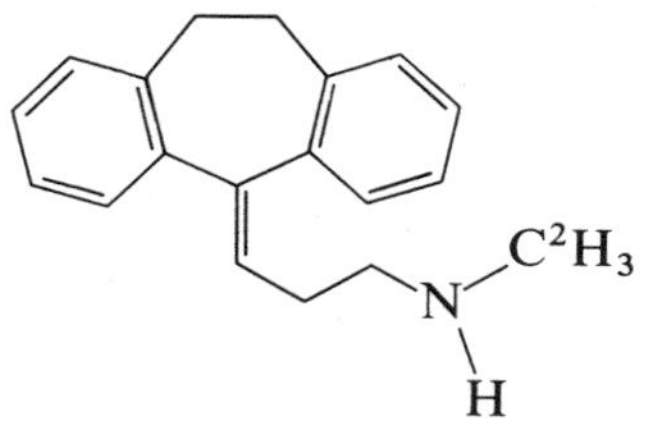

(2)

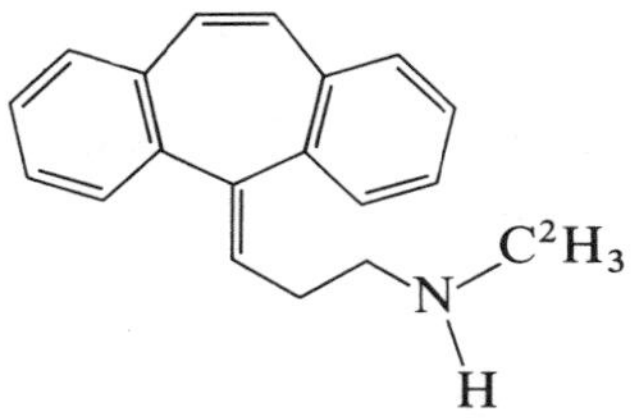

(3)

Consequently a separate experiment involving dosing the labelled compound alone was used to demonstrate an m/e 360 ion at the appropriate g.c. retention time. This precluded the possibility of the observed doublet being due to a fortuitous coincidence with an endogenous contaminant. Another critical question is the number of isotopic atoms to introduce into the molecule. Again it is necessary to balance two conflicting considerations. For g.c./m.s. quantification purposes, analytical precision will be improved by increasing the number of isotopic atoms (because the effect of natural levels of stable isotopes (e.g. ^{13}C at 1.11 %) in giving M + 1 peaks in mass spectra will not interfere with ratio measurements). Nevertheless, with increasing numbers of isotopic atoms the dangers of isotope effects increase, thus the primary purpose of the labelling must decide the issue. For isotope cluster experiments, the number of extra a.m.u. is probably immaterial, so one isotopic atom would suffice (^{18}O has the advantage here since it would satisfy both criteria).

Compounds already possessing their own natural stable isotopes (e.g. ^{35}Cl: ^{37}Cl and ^{79}Br: ^{81}Br) pose their own problems. Metabolic dehalogenation of such compounds bearing stable labels can apparently lead to complicated mass spectral patterns which are not immediately recognizable. Similarly, labelling such molecules with a view to quantitative g.c./m.s. would seem to be likely to lead to complications due to the contributions at higher masses already provided by the natural labels. Nevertheless by using an excess of ^{37}Cl-labelled 2,3,7,8-tetrachlorodibenzo-p-dioxin (TCDD, (4)) of 1000 times the residue level as an internal standard, a p.p.t. g.c./m.s. assay for normatopic TCDD has been published on the basis that the M + 8 natural contribution to the internal standard peak was negligible (Harless et al., 1980). In cases where ^{13}C- or ^{15}N-n.m.r. information is desirable, a good case for stable isotope labelling halogenated compounds can be made but for mass spectral purposes alone the expense does not appear justified.

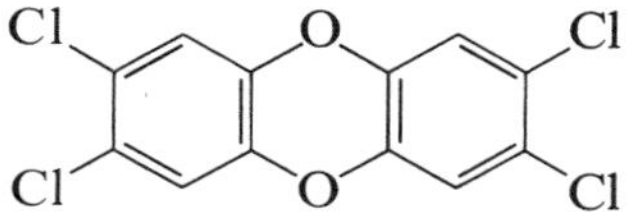

(4)

Admixture with other labels

Some generally applicable questions in relation to mixed labels must be considered. Firstly, is the use of admixed radiochemical precluded? If not, it is difficult to see any reason why the superior 'detectability' of ^{14}C-techniques should be foregone. The decisions to be made in this area are the familiar ones of limits of detection and hence specific activity.

Secondly, we have to consider the ratio in which the normatopic and isotopic chemicals should be mixed to generate ion clusters. The over-all aims of the experiment are the determining factors; where enriched n.m.r. techniques or isotope dilution with normatope are going to be critical, the proportion of isotopic material must be maximized. However, this requirement must be balanced against the potential of the ion cluster technique; if the proportion of one of the components of the doublet is too low, the easy recognition of doublets in even slightly impure mass spectra will be compromised. This difficulty is particularly important when working with traces of metabolites on g.c./m.s./computer systems where there is a 'threshold' effect which can seriously distort peak ratios. To avoid wasting storage space on electronic noise, data systems are often run with a noise rejection threshold. Figure 1 shows (in exaggerated form) the effect of a high threshold on signals in the real ratio of 2:1; the apparent ratio becomes about 4:1 in this case (Millard, 1978a). The effect will only be critical with extremely weak ions where the signal-to-noise ratio of the weaker ion is low; Figure 7b shows a real example at m/e 254/257 and 268/271. Consequently it is probably wise to keep within the 3:1 and 1:3 range of

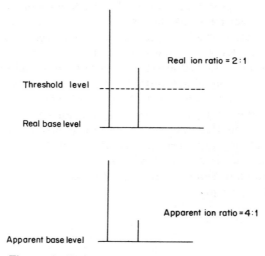

Figure 1 Exaggerated representation of the effect of data system 'thresholds' on apparent ion ratio of weak doublets

mixtures where possible, as the effect becomes more significant with a high ratio of one isotope to the other.

DETECTION AND QUANTIFICATION OF STABLE ISOTOPES

Isotope ratio measuring mass spectrometry

Essentially this technique relies on the conversion of the sample into a gas, e.g. carbon dioxide, whose ratio of ^{13}C to ^{12}C is measured by a fairly simple mass spectrometer system designed to measure simple gas molecular weights to high precision (e.g. 0.001 atom-% excess). The first references to the use of stable isotopes in metabolism studies involved isotope ratio mass measurement of animal excreta following dosing with ^{15}N- or ^{13}C-labelled drugs (Maynert and van Dyke, 1950a, 1950b; Maynert, 1960, 1965; see also von Unruh et al., 1974). In these studies, there were problems due to the low atom % excesses and the difficulties inherent in preparing the required nitrogen or carbon dioxide samples. The technique would be ideal for measuring total recoveries from experiments where no radiolabel had been used. However, it is destructive (sample size 20 μg–10 mg) and yields no information on the structure of the compounds present. Therefore, apart from ^{15}N-applications, it does not seem to offer much to the pesticide chemist who can normally use radiolabels.

Mass spectrometry

This is the conventional means of stable isotope detection and quantification; the examples cited later will show many of the potential uses of the approach. The technique is destructive but as only minute samples are required (e.g. down to the fg level using selective ion monitoring (Harless et al., 1980)) this is not a critical factor. Unfortunately most samples require considerable processing prior to mass spectrometry. Without prior knowledge of the properties of the metabolites, this may only be feasible with radiochemical to ensure that significant metabolites are not lost during 'clean-up'. However, the ease of distinguishing characteristic ion clusters or doublets from the singlets due to endogenous contaminants enables the number of 'clean-up' steps to be minimized with the advantage that the opportunities for forming artefacts are reduced.

Even with the sophisticated derivatization and purification techniques available today, it is often difficult to obtain and/or see molecular ions using conventional electron impact ionization m.s. Thus the use of a multitude of 'soft' ionization techniques has increased recently. They are particularly useful in alliance with stable isotope methodology since the more important higher molecular weight ions are enhanced, thus yielding more visible doublet ions. This is especially true where the label is in a less than ideal position from which it is lost by fragmentation e.g. nortryptyline (see p. 41). Equally important is the effect of reducing the number of fragment ions due to the endogenous contaminants.

Positive and negative ion chemical ionization (CI) is the most used soft-ionization technique and, for the reasons given above, CI spectra may be run on relatively crude samples. Typically the sample is applied to the probe which is very slowly heated to fractionally distil the components of the mixture. By this means, a crude ethylene dichloride extract of the equivalent of 4 ml of urine from a man dosed with 55 mg of a 1:1 mixture of normatopic (5, R = H) and ^{2}H-labelled (5, R = ^{2}H) warfarin was introduced to a mass spectrometer (Pohl et al., 1975). The resulting series of mass spectra were carefully examined for doublet clusters and, as a result, the five or six known rat metabolites (6) were

$[R_{1,2,3,4\,or\,5} = OH, X = O]$
$[R_{1,2,3,4,5} = H, X = H, OH]$

(6)

confirmed without using ^{14}C or time consuming clean-up procedures. Other new soft ionization techniques, e.g. field ionization (FI), field desorption (FD) and desorption chemical ionization (DCI) and their combined use with collision-induced decomposition (CID), permit direct mass spectrometry on molecules traditionally considered to be totally unsuited for the technique. This is likely to be of great importance in stable isotope methodology. A dramatic example has been described for the direct analysis of dinitrophenol in river sludge (Hunt et al., 1980). A 5 mg sample of freeze-dried sludge was inserted on the probe of a triple quadrupole mass spectrometer in the negative ion CI/CID mode. Ions corresponding to the molecular ion of 2,4-dinitrophenol (7) were present at all probe temperatures due to contaminants. However, by inducing fragmentation of this selected ion in a separate chamber of the instrument and looking for the characteristic fragment ions (8) and (9), Hunt and coworkers were able to directly quantify the compounds down to less than 100 ppb. The application of this technique to stable isotope metabolism problems is obvious. Another potential opportunity offered by g.c./m.s./computer systems is the possibility of programming to automatically search g.c. output for characteristic doublet patterns. This can provide a very sensitive metabolite detection system analogous to the g.c./radiogas system (with the advantage of direct quantitation if required).

$m/e = 183$ (7)

$m/e = 153$ (8)

$m/e = 137$ (9)

Nuclear magnetic resonance

This technique is becoming increasingly more valuable to the metabolism chemist; the instrumentation has now improved to the stage where samples of $1-10$ μg are no longer taxing the machines to their limits for proton spectra. Nuclear magnetic resonance is particularly attractive because it is a non-destructive technique. The major problem is that of obtaining adequately pure metabolite samples, although h.p.l.c. now gives a noticeable improvement over t.l.c. techniques. The advantages of h.p.l.c. techniques over t.l.c. are also very noticeable when using field desorption mass spectroscopy (H. R. Schulten, 1980, personal communication). Natural abundance ^{13}C-n.m.r. spectra have always required relatively large amounts of sample due to the combination of its inherent insensitivity relative to the proton (x 0.016) and to its low natural abundance. Nevertheless, current commercial instruments can give satisfactory natural abundance ^{13}C-n.m.r. spectra on $100-200$ μg samples at 25 MHz. ^{13}C-Spectra are less liable to be affected by contamination than ^{1}H-spectra since the range of ^{13}C-signals is ~ 200 ppm. All proton signals occur in a band approximately 10 ppm wide and so the chances of a contaminant peak obscuring a signal of interest is higher in ^{1}H-spectra (e.g. see Rackham, 1980).

^{13}C-Enriched compounds thus offer two major advantages in n.m.r. experiments. Firstly, the enrichment of a sample to $95-100\%$ ^{13}C means that even in a sample that contains only 1% of ^{13}C-labelled compound, 50% of the total ^{13}C-atoms present will be due to that labelled compound. What is more, the rest will be spread over the 200 ppm ^{13}C-range; consequently crude samples can easily be examined (Rueppel et al., 1974, 1977). Secondly, the sample size required to obtain acceptable spectra is much reduced in comparison with natural abundance spectra. An excellent example of the use of n.m.r. to study the metabolism of a ^{13}C-enriched drug is the work on amitryptyline (10) labelled with ^{13}C at the positions shown (Hawkins and Midgley, 1978). The presence of several classes of metabolites containing the intact side chain (10, $R_1 = R_2 = CH_3$) and the mono- (10, $R_1 = H$, $R_2 = CH_3$) and di- (10, $R_1 = R_2 = H$) demethylated side

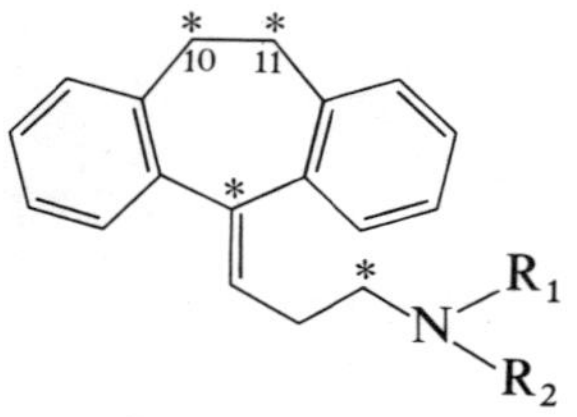

$$R_1 = R_2 = CH_3$$

(10)

chains were deducible from the spectra of crude urine, as was the existence of an N-oxide metabolite. Samples of the order of 800 μg gave adequate spectra with an FT accumulation of only one hour. The results were subsequently confirmed by g.c./m.s. and h.p.l.c.

Nuclear magnetic resonance with other stable label nuclei e.g. ^{15}N is less well documented. However, more and more chemical shift data is becoming available and so the technique has considerable promise. Even where absolute shift data are not well defined, n.m.r. can still give a great deal of information about the relative environment of nuclei of interest. For instance, relaxation time measurements can yield information on enzyme binding sites (e.g. Feeney, 1975; Feeney *et al.*, 1977). ^{15}N-N.m.r. has been used to investigate intact bacterial cell walls (Irving and Lapidot, 1979). Heteronuclear double resonance experiments can give important information on the number, type and exact shift of nuclei coupling with a nucleus of interest (see below).

EXAMPLES OF THE APPLICATION OF STABLE ISOTOPES TO PESTICIDE CHEMISTRY

Since the original ion cluster studies on nortriptyline (Knapp *et al.*, 1972), the technique has been used increasingly for drug metabolism studies (e.g. see the excellent review by Hawkins, 1977). The first application of the doublet ion technique to pesticide metabolism/degradation to appear in print was from a Cyanamid group studying the systemic insecticide cytrolane (mephosfolan, **11**) (Ku *et al.*, 1978, 1979). In these studies, (**11**) labelled with ^{13}C at (α) or (β) was examined under photodegradation and simulated rice paddy conditions. Labels at (α) were introduced from [1-^{13}C]ethanol and those at (β) from potassium-[^{13}C]cyanide. In both cases the compound was admixed with normatopic material in approximately equimolar quantities. For the photodegradative

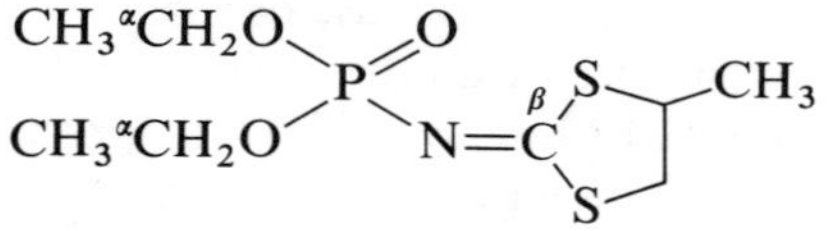

(11)

studies no ^{14}C-label was added, whereas tracer amounts of ^{14}C were included in the paddy field experiment. The group described the use of isotope doublets as 'a means of distinguishing mass peaks due to the metabolites, from the multitude of single peaks from interfering, non-labelled compounds that were always present'. However, apparently only the parent material was identified by the doublet technique since the compound was rapidly degraded to thiocyanate ion and more polar metabolites (some of which were reincorporated into natural rice components). The photodegradation experiments were performed separately on mephosfolan labelled in both positions. Compounds (12A), (13), (14) in addition to (11) were identified by g.c./m.s. (methane CI) in a

chloroform extract of an aqueous solution of (11) which had been irradiated for five weeks with natural sunlight. Subsequent ether extraction of the aqueous fraction (after acidification to pH 2 followed by diazomethylation) and g.c./m.s. (CI) revealed that the acids (12B) and (12C) had been present prior to esterification. The use of the two labelling positions enabled facile identification of these potentially difficult low molecular weight, high polarity degradation products. Particularly useful was the occurrence of $M/M + 2$ doublets when both ethyl groups from the ethyl labelled molecule were present and $M/M + 1$ doublets where one ethyl group had been lost. Gas chromatography peaks which gave singlet ions from ethyl labelled experiments and doublets from imido-^{13}C-label were equally characteristic of ring-derived components. Apparently the use of ion clusters was particularly useful in a photodegradation study performed in natural paddy water which contained many volatile natural products.

In our own laboratories, a study to assess the potential of the isotope cluster technique and other aspects of stable labelling was carried out. The systemic insecticide pirimicarb (15) was chosen as the model compound since it gives a group of minor metabolites that had proved intractable in previous studies with radiolabel alone. The primary metabolic pathways of pirimicarb in plants and animals involve decarbamoylation and 2-N-demethylation. In animals, decarbamoylation proceeds rapidly and thus an experiment was designed to administer stable isotope labelled hydroxypyrimidine (16) to rats. The major problem experienced in the previous studies investigating the minor metabolites

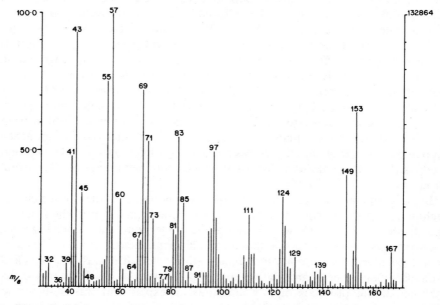

(15)   (16)

of pirimicarb was that of mass spectral interpretation. Even with optimum handling, chromatography and g.c./m.s./computer techniques, spectra similar to that reproduced in Figure 2 were often obtained where trace metabolites were investigated. With hindsight and the availability of reference spectra of synthetic metabolites, peaks of interest can be resolved from those due to contaminants. In the absence of such assistance, much time was wasted in trying to interpret or explain irrelevant ions. The initial problem was how and where to position a label in the hydroxypyrimidine (16). We anticipated doing some quantification studies and therefore we decided to introduce more than one additional mass unit. A ^{14}C-label was normally introduced at carbon 2 of (16) using N^1,N^1-dimethylguanidine as the presursor. Consequently, to use already established synthetic techniques, $[^{13}C,^{15}N_2]$cyanamide labelled at 96.8

Figure 2   Mass spectrometry probe spectrum of unlabelled compound (17) showing the difficulty of distinguishing peaks of interest

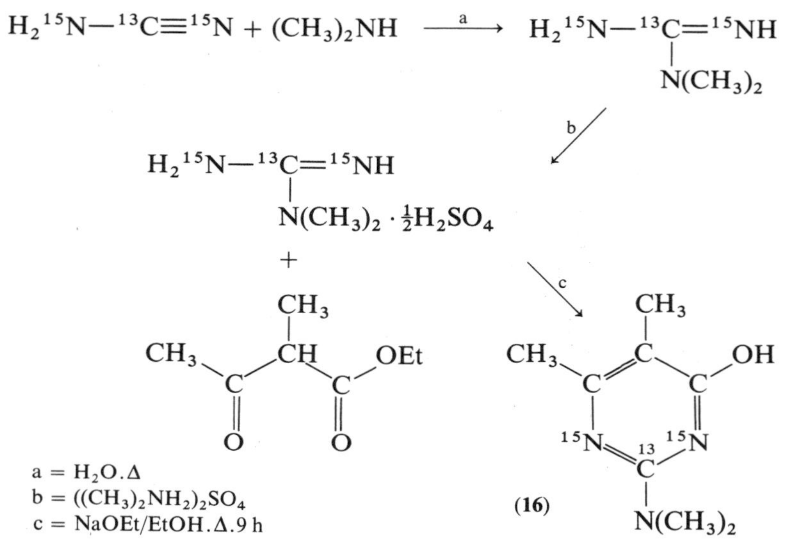

$$H_2{}^{15}N-{}^{13}C\equiv{}^{15}N + (CH_3)_2NH \xrightarrow{\ a\ } H_2{}^{15}N-{}^{13}C={}^{15}NH$$

$$\underset{N(CH_3)_2}{|}$$

$$H_2{}^{15}N-{}^{13}C={}^{15}NH$$
$$\underset{N(CH_3)_2 \cdot \frac{1}{2}H_2SO_4}{|}$$

+

a = H₂O.Δ
b = ((CH₃)₂NH₂)₂SO₄
c = NaOEt/EtOH.Δ.9 h

(16)

Figure 3 The synthetic route used to prepare 2-^{13}C-,1,3-^{15}N$_2$-hydroxy-pyrimidine **(16)**

atom-% ^{13}C and 99.1 atom-% ^{15}N (Prochem, BOC Ltd) was used as the starting material. Figure 3 shows how the cyanamide was converted initially to N^1,N^1-dimethylguanidine and subsequently built up into compound **(16)** with an over-all yield in the order of 50% based on starting material. The methane CI mass spectrum of the synthetic material (MH$^+$, $m/e = 171$) is shown in Figure 4a; no significant contribution from the normatopic material (MH$^+$, $m/e = 168$) is visible.

Since the object of the experiment was to provide information on all aspects of stable isotope labelling, including n.m.r., the isotope was mixed with norma-topic **(16)** in the ratio of 5:2. Approximately 2% by weight of 2-^{14}C-**(16)** (specific activity 50 mCi. mmole^{-1}) was added. The mass spectrum (EI) of the resulting **(16)** is shown in Figure 4b. Compound **(16)** was then administered to rats whose excreta was collected for 48 h. Figure 5 shows how the urine was processed and fractionated; Figure 6 shows a probe m.s. (EI) on a crude sample. The level of contamination is high (note × 10 magnification) but the doublets (m/e, 197/200 and 182/185) make the presence of a metabolite of compound **(16)** obvious and allow the ions at 205, 223 and 279 to be discounted as interferences due to contaminants. There were at least twelve components in the XAD-2 extract; three of these were the well known compounds **(16)**–**(18)** (with traces of their conjugates). Four further trace metabolites were tentatively identified as **(19)**–**(22)**; the ion clusters proved invaluable in the selection of spectra of interest and in mass spectral interpretation.

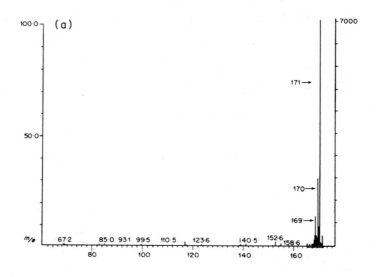

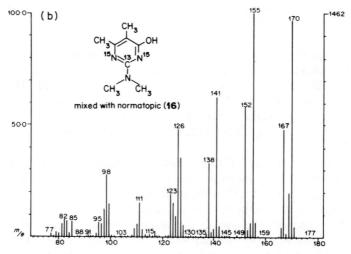

Figure 4 (a) Mass spectrum (CI) of ^{13}C-, $^{15}N_2$-hydroxypyrimidine (**16**), (b) Mass spectrum (EI) of compound (**16**) as used in the isotope cluster experiment

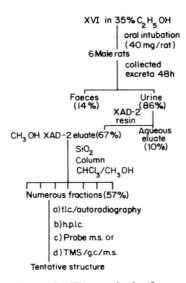

Figure 5 The method of processing rat urine containing labelled (**16**)

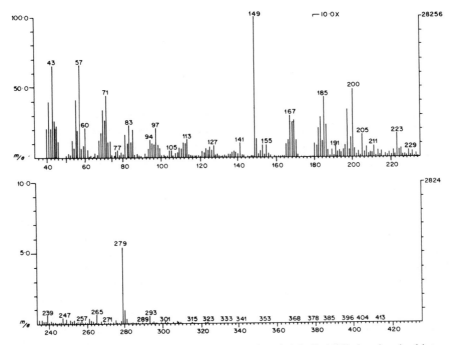

Figure 6 Mass spectrometry probe spectrum of crude labelled (**20**) showing doublets and contaminating background peaks

Compound (19) was readily converted into compound (20) in methanolic solutions; a similar effect has been reported in the literature (Tanaka *et al.*, 1972). The facility of the process was demonstrated by treating a small sample of (20) with ethanol and comparing the mass spectra of their TMS derivatives (Figures 7a and 7b). The characteristic doublet at m/e 283/286 shows that the methyl ether (20) readily exchanged alkyl groups. Figure 7b is also of interest in that the distorted doublets at m/e 254/257 and 268/271 demonstrate the computer 'threshold' effect referred to earlier. This effect was particularly noticeable because of the very small amount of metabolite available.

It was apparent that the ion cluster technique was a valuable complement to ^{14}C in metabolism studies; nevertheless our savings of work-up time were not significant *prior* to the m.s. interpretation stage. Accordingly we investigated the direct derivatization of the crude XAD-2 extract followed by g.c./m.s. The reconstituted ion chromatogram (RIC) was searched for the characteristic doublets by manual inspection of eack peak. The spectra of interest were then cleaned-up by background subtraction and data system enhancement. All the compounds previously identified from chromatographic separation and

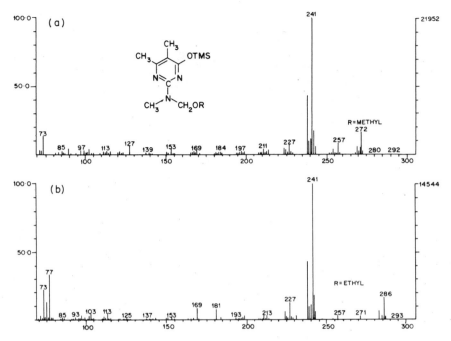

Figure 7   Mass spectrum of mono-TMS ethers of (a) compound (**20**), (b) compound (**20**) treated with ethanol

individual spectroscopy were found and adequate spectra were obtained. Approximately ten additional spectra were obtained containing the characteristic doublets. These were present at only 0.1 % to 0.3 % of the total excreted metabolites and no evidence for them had been obtained on g.c./radiodetection. Thus the direct derivatisation of crude extracts and subsequent g.c./m.s. using doublets to identify peaks of interest was more sensitive than g.c./radiodetection (over a wide concentration range) with the added advantage of avoiding the difficulties of comparing RIC and g.c./radiogas traces. Our next objective was to cut down the time required to search a complete RIC manually for ion clusters. One approach was to predict the masses of derivatized potential metabolites and to use the computer to search the RIC for the appropriate pairs of ions. Figure 8 shows the results of such a search; the simultaneous maximization of the relevant pairs is clearly seen; the bottom trace is the RIC. Unfortunately, this technique will never pick out unexpected metabolites and only has real potential for demonstrating the presence or absence of a particular compound of interest in an extract.

The final approach was more comprehensive. We set out to design a computer search program to locate P/P + 3 doublets at any mass between 100 and

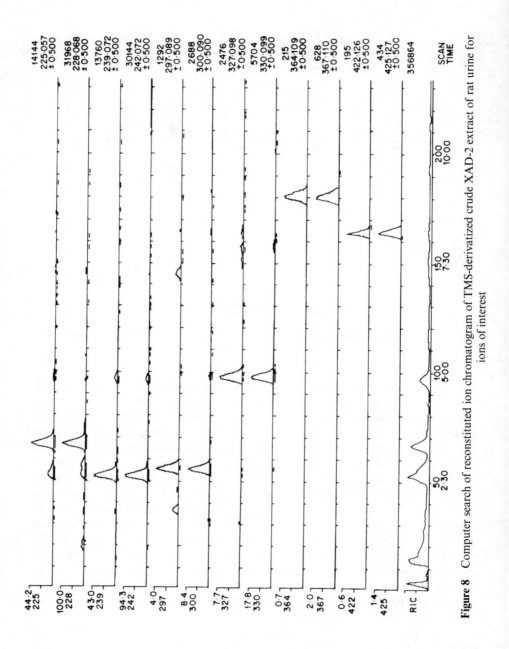

Figure 8 Computer search of reconstituted ion chromatogram of TMS-derivatized crude XAD-2 extract of rat urine for ions of interest

450 a.m.u. There is considerable scope in this area and a program is available to provide a selective g.c./m.s./computer detector for polyhalogenated pesticides (Canada and Regnier, 1976). In our program, the RIC data is initially refined by a Biller–Bieman type analysis to reduce the number of scans to be searched. Each enhanced scan is compared with a specially created library of doublet ions where each entry contains only one P/P + 3 doublet in the appropriate ratio. There are library entries corresponding to every m/e value from 100/103 to 450/453. The mass spectra of those scans which give a significant

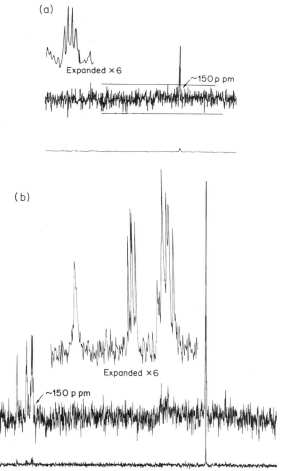

Figure 9 ^{13}C-n.m.r. Spectra of (a) 100 μg of (**16**) in crude plant extract, (b) approximately 1 mg equivalent of metabolites of compound (**16**) in filtered raw rat urine

fit with the library are then automatically printed out: simple manual scanning of these finally eliminates those containing merely fortuitous doublets. By this means an RIC containing 250 scans was automatically reduced to 24 scans of interest; a further eight scans were rejectable by eye to leave the same set of spectra containing characteristic doublets due to compound (16) metabolites that had taken several hours to obtain by manual searching.

Figure 9a shows the n.m.r. spectrum of 100 μg of compound (16) in a crude plant extract (15 h accumulation). The expanded signal shows the effect of coupling with the adjacent ^{15}N nuclei (this caused a reduction in the achievable signal-to-noise ratio for the ^{13}C-label in this experiment). Figure 9b shows the ^{13}C-n.m.r. spectrum of filtered rat urine containing a total of 1 mg equivalent of metabolites of compound (16). This demonstrates at least three major metabolites. However, their chemical shifts are very similar, probably indicating that no metabolism occurred directly at C-2. No further work has been done on these n.m.r. spectra, but the potential for obtaining information from very crude extracts is clear.

There is one published example of the very effective use of ^{13}C-n.m.r. in the pesticide field (Rueppel et al., 1977). In this study of the metabolism of ^{14}C- and ^{13}C-labelled glyphosate (23) in soil, it was found that parent (23) and the amino compound (24) formed the major components of soil aqueous extracts. The

$$\underset{(23)}{HOOCCH_2NH\overset{*}{C}H_2\overset{\displaystyle O}{\overset{\displaystyle \|}{\underset{\displaystyle OH}{P}}}{-}OH} \qquad \underset{(24)}{H_2NCH_2\overset{\displaystyle O}{\overset{\displaystyle \|}{\underset{\displaystyle OH}{P}}}{-}OH}$$

degradation of 25 mg of N-phosphono[^{13}C]methyl glycine (90 atom-% ^{13}C at position * in structure (23)) containing a trace of ^{14}C-material in an aerobic soil shake-flask culture was studied. The aqueous supernatant after centrifugation was concentrated and filtered prior to measuring the n.m.r. spectrum; the equivalent of 7–8 mg of material was present at this stage (distributed approximately equally between (23) and (24)). The n.m.r. spectrum showed signals characteristic of the labelled carbon atoms of the two compounds clearly resolved from one another and of approximately equal intensities. Despite the fact that the ^{13}C signals appeared as doublets (due to phosphorus–carbon coupling) the signal to noise ratio was very acceptable and presumably, with the current generation of spectrometers, considerable reduction in sample size could be expected today. Most importantly, despite the large amounts of dissolved solids present in aqueous soil extracts, the shift values of the ^{13}C-atoms were not changed significantly from those of the pure standards. Thus in this case where major changes in the structure of the molecule have occurred near the labelled centre, ^{13}C-n.m.r. provided a great deal of information without the need for significant clean-up steps. In the same paper, the Monsanto group made use of the pseudo-INDOR type technique of locating the chemical shift of the ^{31}P-atom of trace samples. The ^{31}P–^{1}H doublet corresponding to the phosphonomethylene –CH$_2$– in (23) in the more sensitive ^{1}H spectrum was examined while scanning the ^{31}P region. The shift value for the ^{31}P atom was located when the –CH$_2$– doublet collapsed to a singlet. This may well prove a useful technique in some stable isotope labelling applications. In several presentations (Rueppel *et al.*, 1975a, 1975b, 1979) the same group have illustrated the utility of ^{13}C-labelling to facilitate soil, water and plant metabolism studies. A particularly interesting technique was their adaption of isotope dilution to reduce the handling problems associated with the derivatization and manipulation of low μg amounts of metabolite. From an experiment using only ^{13}C-*N*-phosphonomethyl glycine with a trace of ^{14}C material (i.e. no normatopic compound) the team isolated a plant extract containing approximately 20 μg of a metabolite believed to be compound (24). To simplify the identification, 100 μg of the putative unlabelled metabolite was added and the sample was derivatized and its mass spectrum obtained. The characteristic doublet peaks confirmed the identity of the plant metabolite and the amount present in the plant extract could be calculated from the relative peak height ratios in the mass spectrum.

A final, unusual, example of the application of the doublet ion cluster technique comes from the use of the prostaglandin analogue cloprostenol (25) in

(25)

the veterinary field. Because the label was ^{14}C at high specific activity (approximately 30 mCi. mmole^{-1}) approximately equivalent P and P + 2 ion signals were seen in the metabolite mass spectra. The authors claimed that the clusters were 'analogous to those often obtained with stable isotopes' (Bourne *et al.*, 1979, 1980). When concentrating and isolating such labelled compounds, great care is presumably necessary to avoid the problems associated with autoradiolysis.

OTHER POSSIBLE APPLICATIONS OF STABLE ISOTOPE LABELLING TO PESTICIDE CHEMISTRY

So far we have largely considered the isotope cluster aspect of stable isotope labelling in detail. There are many other potential applications but the majority of the examples come from fields other than pesticide chemistry.

Quantification

Much has been written on the application of stable isotopes to analytical methodology; both as carriers and as internal standards (e.g. Hawkins, 1977; Millard, 1978a). While the finer points of accuracy, precision and mass spectrometer stability in relation to the enhanced ease of analysis are arguable (e.g. Picart et al., 1978; Millard, 1978b); it is indisputable that stable isotope analogues, provided they are of sufficient stability and purity, are valuable assets to the analyst. This is particularly true where the number of samples likely to be analysed by the method is small enough to justify the use of expensive mass spectrometer time. A frequent example in pesticide chemistry is the analysis for pesticide metabolites in tissues of animals fed trace levels of pesticide to simulate the ingestion of treated fodder. For a routine method generating many results over a long period, it is normally worth investing more method development time to provide samples amenable to g.c./specific detection in order to avoid monopolizing a mass spectrometer.

An example of the quantitative use of stable isotopes in pesticide chemistry is an analysis for paraquat (**26**), in marihuana pyrolysis products (Wall et al., 1979) using an N-methyl-hexadeutero-analogue. This method gave linear results between 0.01 ppm and 10 ppm of paraquat, although no exhaustive investigation of accuracy and reproducibility was carried out.

$$CH_3-\overset{+}{N} \text{=}\!\!\text{=} \overset{+}{N}-CH_3 \qquad Cl^- \qquad Cl^-$$

(26)

$$O_2N \text{...} \begin{matrix} N \\ N \end{matrix} \text{...} \begin{matrix} CH_3 \\ R \end{matrix} \qquad CH_3\ CH_3$$

(27)

The turkey anti-histomoniasis compound, ipronidazole (IP, **27**, R = H) and its hydroxylated metabolite (HIP, **27**, R = OH) have been monitored by g.c./negative ion CI m.s. using the N-methyl-$^2H_3$-stable isotope labelled compounds as internal standards (Garland et al., 1980). The addition of IP-$^2H_3$ and HIP-$^2H_3$ (200 ng of each) to all samples followed by clean-up and g.c./m.s. gave detection at the 0.2 ng level with a signal-to-noise ratio greater than 50:1. Negative ion CI mass spectrometry was selected because it afforded a 20-fold increase in ion current when compared to EI, thus giving greater sensitivity. Additional specificity arose from the fact that few natural products ionize under negative chemical ionization conditions.

Other relevant analyses include that of the plant hormone, indole-3-acetic acid (IAA) which has been monitored in plant tissues at the 2 ppm level using $^2H_2$-label in the acid methylene group (Caruso et al., 1978). Similarly, ^{18}O has been used to measure pentachlorophenol levels at the 1 ppb level in water (Ingram et al., 1979). The general references cited earlier will give access to the much more extensive pharmaceutical literature in this field. There it becomes apparent that isotopes such as ^{13}C and ^{15}N are definitely to be preferred to ^2H since the dangers of isotope effects are minimized.

Isotope dilution

The above were largely examples of direct isotope dilution: the inverse of this is the reverse isotope dilution method traditionally used by metabolism chemists. This technique may suffer from problems caused by physical entrapment and cocrystallization which can confuse studies using ^{14}C measurements. Related labelled compounds may repeatedly cocrystallize with unlabelled standard to give erroneously high (but apparently constant) specific activities. However, the specificity of mass spectrometric measurement avoids this pitfall. The example of compound (24) already quoted demonstrates how stable isotopes and a mass spectrometer can avoid these problems. Various modifications of the technique can be devised to solve individual problems. The technique has also been used to quantify the uptake and disappearance of trifluralin (TFN, 28) in rat tissues (Heck *et al.*, 1977). Rats were dosed with $^2H_7$-TFN and samples were collected and stored deep frozen. For analysis, unlabelled TFN (80 μg) was added to each sample and the TFN was then purified by extraction, g.p.c., h.p.l.c. and a final extraction. The $^2H_0$ and $^2H_7$ species were monitored by field ionization mass spectrometry giving a minimum detectable concentration of 0.5 ppb and a method linear over three orders of magnitude. The possibility of isotope effects in the metabolism of the $^2H_7$-material suggests that dosing with $^2H_0$-TFN and isotope dilution with the $^2H_7$-labelled chemical might have given more realistic pharmacokinetic data.

In an interesting study on the metabolism of the central muscle relaxant,

$$CH_3(CH_2)_2\diagdown N \diagup (C^2H_2)_2C^2H_3$$
$$O_2N\diagdown \qquad \diagup NO_2$$
$$CF_3$$
(28)

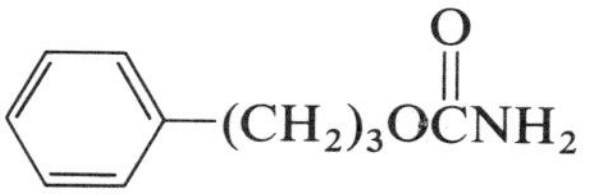

(29)

(30)

3-phenyl propylcarbamate (29), in rats and man, both $^2H_5$-(arom) and 3-^{13}C-labels were used (Horie and Baba, 1979). By means of various isotope dilution experiments, eight metabolites were quantified. This study was additionally complicated because one of the metabolites, hippuric acid (30), is also present as an endogenous product in urine.

Biosynthetic aspects

The increasingly important role of stable isotopes in biosynthetic studies highlights a potential for problem solving in an area of great interest to the pesticide biochemist. In these experiments (e.g. Leete, 1977 and other papers), ^{13}C-labelled putative precursors to natural plant secondary metabolites are administered to plants. The isolated natural product can be examined by ^{13}C-n.m.r. to investigate immediately and directly which carbon atoms have been labelled. In the past, ^{14}C label was used for this purpose but sophisticated and time consuming degradative chemistry was essential to identify the positions of the carbon atom(s) bearing the label. Now, ^{14}C is often added with the ^{13}C-precursors but normally only to assist in the isolation of the natural products and the quantification of the incorporation. The use of two adjacent ^{13}C-atoms in the precursor is now commonplace; this gives rise to ^{13}C–^{13}C-n.m.r. coupling to give a doublet located on both sides of the natural abundance chemical shift position of the carbon of interest (e.g. see Rackham, 1980). This has the two-fold advantage of enabling the detection of lower specific incorporations and of determining whether the ^{13}C–^{13}C bond in the precursor had been broken prior to incorporation (since a doublet would no longer appear in the natural product). This technique would seem to have application to the thorny problem of 'strongly bound' residues in plants and soil (e.g. Still et al., 1976, 1979). By using a minimum of ^{14}C-label as a tracer to locate and isolate pesticide derived 'bound' fractions, the bulk of the added pesticide could be stable labelled in such a way as to give n.m.r. spectra of value in understanding the mode of binding of the residue. Similar experiments could be envisaged to provide direct information on the complete breakdown of pesticides and the subsequent anabolism of the fragments into natural products (e.g. Ku et al., 1978). Support for this potential application comes from the measurement of ^{15}N-enrichment of amino acids in the order of 0.08 atom-% (Matthews et al., 1979). Similarly 1-^{13}C-1, 1-^{2}H-ethanol has been used to establish which carbon atoms in cholic acid (31) were derived from ethanol (Cronholm et al., 1975).

One interesting example of the application of stable isotopes to natural products associated with pesticides has already been reported (More et al., 1978). An investigation of the conjugation of 3-phenoxybenzoic acid (32) by excised shoots and leaves of various plants identified the glycosidic conjugate (33); a further conjugation of (33) with an arabinose unit to give a disaccharide compound was seen. By maintaining abscised leaves in an atmosphere con-

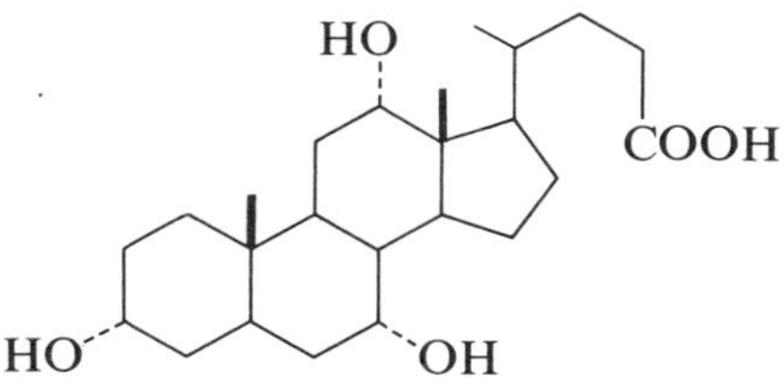

(31)

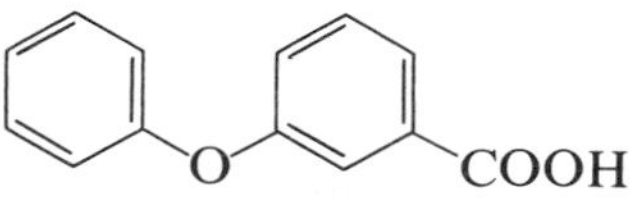

(32)

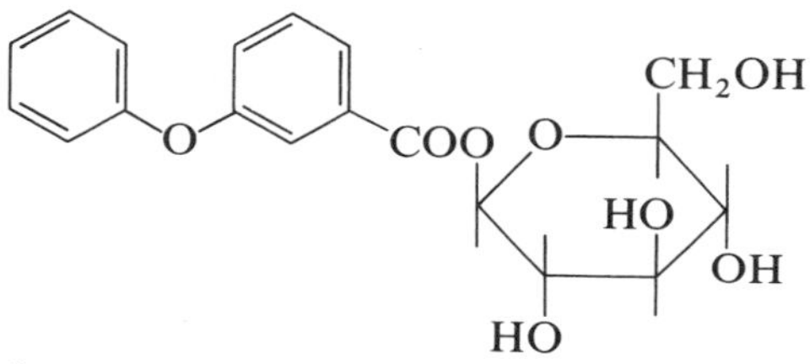

(33)

taining $^{13}CO_2$ for 2 h and then adding $3\text{-}^{14}C$-labelled (32) to the leaf nutrient, an extract was ultimately obtained containing (33). Subsequent acetylation and mass spectrometry showed a cluster of molecular ions between m/e 544 and 550. Natural and synthetic (33) acetate has an ion of m/e 544 and hence the conclusion was that (32) undergoes conjugation with at least some *recently formed* glucose.

Derivatization labelling

So far, discussion has centred around the stable isotope labelling of the parent pesticide; an alternative is the use of stable isotopically labelled derivatization reagents. A mixture of normatopic and isotopic reagents will result in derivatives bearing doublet ions in the mass spectrum and their occurrence in m.s. fragments can help to elucidate sites of metabolism. Similarly the occurrence of singlets in a g.c./m.s. trace of such derivatized metabolites allows non-derivatized endogenous molecules to be ignored. The technique also has particular relevance in *N*-methyl pesticides which undergo *N*-demethylation on metabolism. For example, permethylation of a crude metabolite mixture containing compounds (16)–(18) with $[C^2H_3]$methyl iodide leads to the two compounds (34) and (35)

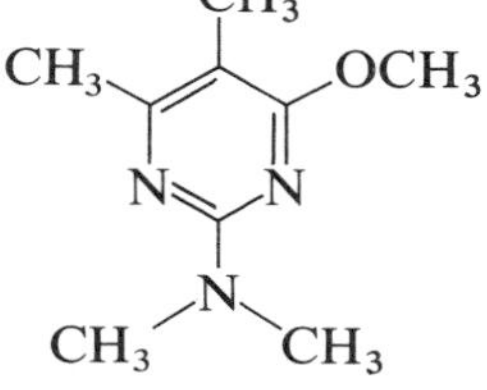

(34)

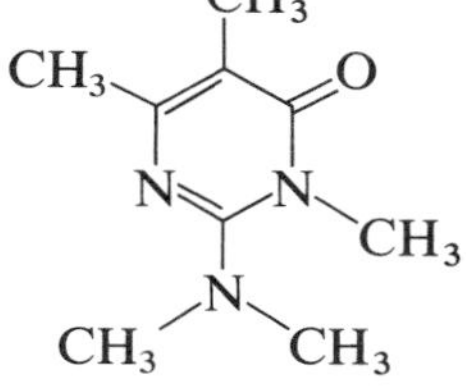

(35)

(arising from the ketoenol nature of the hydroxypyrimidine ring system) giving a series of molecular ions corresponding to the numbers of methyl groups lost in metabolism and permitting a measurement of the relative levels of the three compounds. The alternative technique of alkylation with another group e.g. ethyl, gives rise to mixtures of derivatives each having its own retention time on g.c. Again the use of stable labels has simplified a typical problem. Another potential application of derivatization labelling is in the derivatization with ion clusters of classes of natural compounds to facilitate their profiling. This could be useful in mode of action studies where the concentrations of various fungal sterols were being monitored. Hunt and coworkers have used the formation of derivatization doublets of peptides by mixtures of CH_3- and $[C^2H_3]$acetic anhydrides followed by methylation to facilitate the identification of N-terminal amino acids in the mass spectral determination of peptide structures (Hunt et al., 1980). Some derivatives give rise to characteristic m.s. fragmentation patterns; e.g. trimethylsilyl(TMS) derivatives give molecular ions and $M - 15$ peaks. Thus the use of mixtures of normatopic and isotopic TMS reagents coupled with a computerized search procedure for linked $P/P + n$ and $P - 15/(P + n) - 15$ ion pairs would give highly sensitive m.s. detection of TMS derivatized molecules in a g.c./m.s. scan.

Toxicological and mode of action studies

Although the problem of avoiding ^{14}C radiation burden to humans is not normally relevant to the pesticide toxicologist, the ability of the mass spectrometer to be specific for parent molecules in the presence of metabolites does offer some attraction. No examples have been reported in the pesticide field as yet.

The SICAT (Stable Isotope Co-administration Technique) method is a pharmacokinetic tool used to compare bioavailability via different routes of administration (e.g. Haskins et al., 1980a,b). In this method the accumulation, disposition and dissipation of a compound entering the body orally and intravenously can be measured simultaneously by dosing with unlabelled compound via one route and stable isotope labelled parent via the other. Time is saved because no 'wash-out' period is required, with the added advantage that the potential pitfalls of changes in metabolism etc are avoided. Moreover, the number of sets of samples that have to be analyzed is halved. Care is necessary in the interpretation of data from all such studies to guard against the possibilities of isotope effects or of mass spectral impurities in one or other dosing solution. This technique may have potential in formulation, mode of action or field studies of pesticides. An example is seen with the drug benoxaprofen (**36**); two formulations—one of large and the other of small crystal sizes—were administered simultaneously, one was the normatopic material and the other was stable labelled. The more efficient formulation was determinable by g.c./m.s. measurement of blood levels (Wolen et al., 1979).

(36)

$$FCH_2 \overset{\overset{\displaystyle NH_2}{|}}{\underset{\underset{\displaystyle R}{|}}{C}} COOH \qquad CH_3CH_2O \text{—} \langle\text{—}\rangle \text{—} NHCOCH_3$$

(37)   (R = H or ^{2}H)   (38)

A related important use is pulse labelling to investigate the pharmacokinetics of a compound when a subject has reached a steady state concentration (e.g. Kapetanovic and Kupferberg, 1980). Outside the toxicological sphere, this technique might find application in field studies where the widespread use of radiochemicals is normally impracticable. The toxicological and mode of action aspects are interrelated in the use of stable isotope labelling to provide the chemist with a 'pseudo-racemate'. In other words, a racemic mixture where one enantiomer is labelled with isotope and the other is normatopic. The relative metabolism and pharmacokinetics of such pseudo-racemates have been examined (e.g. Hachey et al., 1979; Weinkam et al., 1976; Jarman et al., 1979). Where the optical antipodes have been shown to exhibit different biological properties, great care must be taken to avoid isotope effects since the relevant enzyme system is obviously susceptible to subtle changes. ^{2}H-Labelling at the optical centre is particularly unwise and indeed this has been shown to be an effective means of slowing down the metabolism of an amino acid (37) and hence increasing its antibacterial half-life (see Draffan, 1978).

The enantiomers of several pesticides (e.g. the pyrethroid decamethrin) have been shown to demonstrate different biological activities and so stable isotopes present another tool with which to examine these effects. ^{14}C-Labelling would not give sufficiently precise answers here since metabolite levels would also be measured. Pseudo-racemates also provide a relatively simple means of measuring enantiomeric ratios at the submicrogram level (e.g. Howald et al., 1980).

Mode of action studies themselves are normally wholly individual, aimed at prediction of more active analogues or explaining a particular problem relating to registration or toxicology. An example of how stable isotope might be applied is the use of ^{18}O-label to identify the source of oxygen for metabolism. This has been used to establish whether epoxide intermediates were significant in the metabolism of phenacetin (38) (Nelson et al., 1979). Similar problems of the covalent binding of reactive intermediates to macromolecules could also be tackled via stable isotope labelling using both n.m.r. and mass spectrometric techniques.

(39)

The imaginative research into the metabolism of clonidine (39) using the labels shown, embodies many of the topics previously discussed (see Figure 10). The ^{35}Cl: ^{37}Cl doublets were used as ion cluster generators to pick out g.c./m.s. peaks of interest. The ^{2}H-labels were used to determine the sites of metabolic action while the $^{13}C_2$-labelled compound served to show that glyoxal and glyoxalate were formed by preparing the quinoxoline derivatives and examining their mass spectra. The metabolites shown were confirmed and quantified by reverse isotope dilution with unlabelled metabolite on the extracts of rat liver microsomes challenged with the appropriately labelled clonidine. Some of the stable labelled compounds were also used as conventional internal standards in assays on unlabelled clonidine extracts. Additionally, the number of methylatable sites on metabolites was determined by comparing the masses of methyl derivatives prepared by on-column methylation with trimethylanilinium hydrochloride or its $^{2}H_9$-analogue. The mechanism of imidazoline ring oxidation was investigated by using an atmosphere of $^{18}O_2$. Despite the comprehensive and successful application of stable isotopes, the authors conclude that ^{14}C-labels were essential for the easy detection of metabolites (Davies *et al.*, 1979; Baillee *et al.*, 1979).

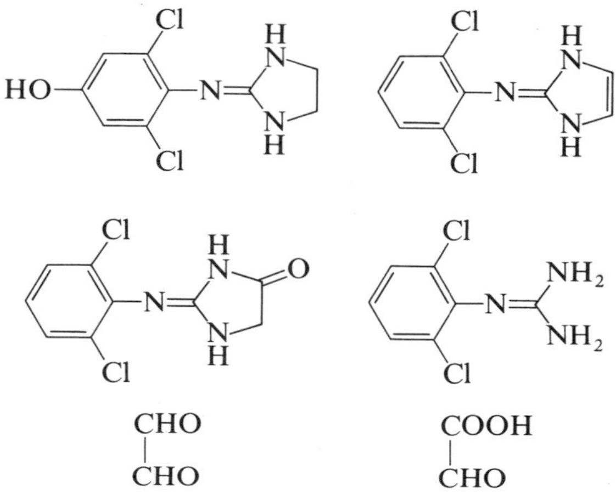

Figure 10   The metabolites of clonidine

CONCLUSIONS

This chapter set out to outline some of the factors to consider when planning stable isotope studies; to describe stable isotope labelling applications already in the pesticide literature and, probably most importantly, to suggest areas in which stable isotopes might provide solutions to problems more easily and cheaply than other techniques. The literature review is by no means exhaustive; nevertheless, it provides suitable access into most of the specific areas discussed. Discussions with colleagues in the pesticide industry have revealed that there is increasing interest in the use of stable isotopes and one or two laboratories already seem heavily committed to the technique. Therefore we can hopefully expect the number of published pesticide metabolism and residue analysis studies to increase soon. Publications in the area are particularly important to establish the validity and credibility of the technique (particularly in conjunction with radiolabelling studies) in the eyes of both the scientific and regulatory communities. The critical factor is, of course, the cost. The five main advantages offered by stable isotopes listed in the introduction and exemplified in the subsequent sections must be weighed against the disadvantages. These are the intrinsic cost of preparing the labelled compounds (in addition to the essential radiolabel) allied with the high cost and complexity of the technology required for stable isotope detection. However, it is probably true to say that the mass spectrometer and FT-n.m.r. already form an essential part of the equipment required by the metabolism chemist and so the argument here centres around the instrument and operator time involved. The cost of synthesis may well become less significant with time since, with increasing demand, the relative cost of precursors may well decrease over the next few years and the range of available compounds should increase. The future of the technique lies largely in the hands and inventive abilities of the pesticide chemist. It is also important to stress that the future of both mass and nuclear magnetic resonance spectrometry has a key role to play. Every step forward in the ability to ionise molecules and each improvement in n.m.r. instrument sensitivity will add considerably to the potential of stable isotope labelling. To take an overview, stable isotope labelling does not represent a panacea; like all techniques, it has drawbacks. However, as a complement to the use of radiolabels the application of stable isotopes should speed up pesticide metabolism studies and provide a possible means of solving some of the problems that have proved intractable in the past.

REFERENCES

Baillee, T. A., Neill, E., Hughes, H., Davies, D. L., and Davies, D. S. (1979) 'Application of stable isotope labelling in studies of the metabolism of clonidine in rat liver' in *Stable Isotopes: Proceedings of the Third International Conference* (E. R. Klein and P. D. Klein, Eds.), Academic Press, New York, pp. 415–425.

Blake, M. I., Crespi, H. L., and Katz, J. J. (1975) 'Studies with deuterated drugs', *J. Pharm. Sci.*, **64**, 367–391.

Bourne, G. R., Moss, S. R., Phillips, P. J., Webster, J. T. A., and White, D. F. (1979) 'Ion cluster techniques in drug metabolism: an example of the use of mixtures of the ^{12}C-:^{14}C- isotopic forms of the synthetic prostaglandin cloprostenol ('Estrumate') to facilitate metabolite identification', *Biomed. Mass Spectrom.*, **6**, 359–360.

Bourne, G. R., Moss, S. R., Phillips, P. J., and Shuker, B. (1980) 'The metabolic fate of the synthetic prostaglandin cloprostenol ('Estrumate') in the cow: Use of ion cluster techniques to facilitate metabolite identification', *Biomed. Mass Spectrom.*, **7**, 226–230.

Canada, D. C., and Regnier, F. E. (1976) 'Isotope ratios as a characteristic selection technique for mass chromatography', *J. Chrom. Sci.*, **14**, 149–154.

Caruso, J. L., Smith, R. G., Smith, L. M., Cheng, T. Y., and Davies, G. D., Jr. (1978) 'Determination of indole-3-acetic acid in Douglas fir using a deuterated analogue and selected ion monitoring', *Plant Physiol.*, **62**, 841–845.

Cronholm, T., Sjövall, J., and Burlingame, A. L. (1975) 'Stable isotopes, mass-spectrometry and ^{13}C-n.m.r. in studies of bile acid biosynthesis and metabolism' in *Adv. Bile Acid Res., Bile Acid Meet. 3rd. 1974* (S. Matern, J. Hackenschmidt and P. Back, Eds.), Schattauer, Stuttgart, Germany, pp. 25–30.

Davies, D. S., Baillee, T. A., Neill, E., Hughes, H., and Davies, D. L. (1979) 'Applications of stable isotope labelling in studies of the pharmacokinetics and metabolism of clonidine', *Adv. Pharmacol. Ther., Proc. Int. Congr. Pharmacol.*, **7th, Series 7**, 215–223.

Draffan, G. H. (1978) 'Stable isotopes in human drug metabolism studies' in *Stable Isotopes: Applications in Pharmacology, toxicology and Chemical Research* (T. A. Baillee, Ed.), Macmillan, London, pp. 27–42.

Evans, E. A. (1976) *Self-decomposition of Radiochemicals*, Review 16, The Radiochemical Centre Ltd., Amersham, U.K.

Feeney, J. (1975) 'The application of ^{13}C-n.m.r. spectroscopic techniques to biological problems', in *New Techniques in Biophysics and Cell Biology*, (R. H. Pain and B. J. Smith, Eds.), J. Wiley, Chichester, Vol. 2.

Feeney, J., Birdsall, B., Roberts, G. C. K., Griffiths, D. V., and Burgen, A. S. V. (1977)

'^{1}H-Nuclear magnetic resonance studies of lactobacillus casei dihydrofolate reductase', *Proc. Royal Soc. of London, B*, **196**, 251–265.

Garland, W. A., Hodshon, B. J., Chen, G., Weiss, G., Felicito, N. R., and MacDonald, A. (1980) 'Determination of ipronidazole and its principal metabolite in Turkey skin and muscle by gc/negative chemical ionization mass spectrometry-stable isotope dilution', *J. Agr. Food Chem.*, **28**, 273–277.

Gordon, A. J., and Ford, R. A. (1972) *The Chemists Companion*, John Wiley and Sons, New York, pp. 88–106.

Hachey, D. L., Kreek, M. J., and Klein, P. D. (1979) 'Stereoselective disposition of R-(−)- and S-(+)-methadone in Man' in *Stable Isotopes: Proceedings of the Third International Conference* (E. R. Klein and P. D. Klein, Eds.), Academic Press, New York, pp. 411–414.

Harless, R. L., Oswald, E. O., Wilkinson, M. K., Dupuy, A. E., Jr., McDaniel, D. D., and Tai, H. (1980) 'Sample preparation and gas chromatography-mass spectrometry determination of 2,3,7,8-tetrachloro-dibenzo-p-dioxin', *Anal. Chem.*, **52**, 1239–1245.

Haskins, N. J., Ford, G. C., Palmer, R. F., and Waddell, K. A. (1980a) 'A stable isotope dilution assay for disopyramide and its [^{13}C-,^{15}N-] labelled analogue in biological fluids', *Biomed. Mass Spectrom.*, **7**, 74–79.

Haskins, N. J., Ford, G. C., Waddell, K. A., Spalton, P. N., Walls, C. M., Forrest, T. J., and Palmer, R. F. (1980b). 'The absorption of disopyramide in animals determined using a stable isotope co-administration technique', *Biomed. Mass Spectrom.*, **7**, 80–83.

Hauck, R. D., and Bystrom, M. (1970) ^{15}*N—A selected Bibliography for Agricultural Scientists*, Iowa State University Press, Iowa.

Hauck, R. D., and Bystrom, M. (1981) ^{15}*N—A selected Bibliography for Agricultural Scientists*, Tennessee Valley Authority, (in press).

Hawkins, D. R. (1977) 'The role of stable isotopes in drug metabolism' in *Progress in Drug Metabolism* (J. W. Bridges and L. F. Chasseaud, Eds.), Vol. 2, pp. 163–218. John Wiley, Chichester.

Hawkins, D. R., and Midgley, I. (1978) 'The use of ^{13}C-n.m.r. spectroscopy for the detection and identification of metabolites of carbon-13 labelled amitryptyline', *J. Pharm. Pharmacol.*, **30**, 547–553.

Heck, H. d'A., Dyer, R. L., Scott, A. C., and Anbar, M. (1977) 'Determination and disposition of trifluralin in the rat. Separation by sequential high-pressure liquid chromatography and quantitation by field ionization mass spectrometry', *J. Agr. Food Chem.*, **25**, 901–907.

Horie, M., and Baba, S. (1979) 'Studies on drug metabolism by use of isotopes XXIV—determination of 3-phenylpropyl carbamate metabolites using stable isotope labelling with ^{2}H- or ^{13}C-', *Biomed. Mass Spectrom.*, **6**, 63–66.

Horning, M. G., Nowlin, J., Lertratanangkoon, K., and Thenot, J. P. (1979a) 'The use of stable isotopes in the quantification of drugs', in *Instrum. Appl. Forensic Drug Chem.*, *Proc. Int. Symp. 1978* (Pub. 79) (E. Klein, A. V. Kreugel and S. P. Sobol, Eds.), G.P.O., Washington D.C., pp. 41–47.

Horning, M. G., Thenot, J. P., Bouwsma, O., Nowlin, J., and Lertratanangkoon, K. (1979b) 'Changes in chemical and biological properties of drugs due to deuterium labelling', in *Adv. Pharmacol. Ther.*, *Proc. Int. Congr. Pharmacol.*, *7th*, Pergamon, Oxford, pp. 245–256.

Howald, W. N., Bush, E. D., Trager, W. F., O'Reilly, R. A., and Motley, C. H. (1980) 'A stable isotope assay for pseudoracemic warfarin from human plasma samples', *Biomed. Mass Spectrom.*, **7**, 35–39.

Hunt, D. F., Shabanowitz, J., and Giordani, A. B. (1980) 'Collision activated decompositions of negative ions in mixture analysis with a triple quadrupole mass spectrometer', *Anal. Chem.*, **52**, 386–390.

Ingram, L. L. Jr., McGinnis, G. D., and Parikh, S. V. (1979) 'Determination of penta-chlorophenol in water by mass spectrometric isotope dilution', *Anal. Chem.*, **51**, 1077–1078.

Irving, C. S., and Lapidot, A. (1979). 'Bacterial cell wall motion, an *in vivo* ^{15}N-n.m.r. study', in *Stable Isotopes: Proceedings of the Third International Conference* (E. R. Klein and P. D. Klein, Eds.), Academic Press, New York, pp. 307–315.

Jarman, M., Cox, P. J., Farmer, P. B., Foster, A. B., Milsted, R. A. V., Kinas, R. W., and Stec, W. J. (1979) 'The use of deuteriom labelled analogues in a study of the metabolism of the enantiomers of cyclophosphamide', in *Stable Isotopes: Proceedings of the Third International Conference* (E. R. Klein and P. D. Klein, Eds.), Academic Press, New York, pp. 363–369.

Jarman, M., and Foster, A. B. (1979) 'Metabolism directed design of anticancer Agents: Applications of deuterium labelling' in *Adv. Pharmacol. Ther., Proc. Int. Congr. Pharmacol.*, *7th*, Pergamon, Oxford, pp. 225–233.

Kapetanovic, I. M., and Kupferberg, H. J. (1980) 'Stable isotope methodology and Gas Chromatography Mass Spectroscopy in a Pharmacokinetic study of Phenobarbital', *Biomed. Mass Spectrom.*, **7**, 47–52.

Klein, E. R., and Klein, P. D. (1978) 'A selected bibliography of biomedical and environmental applications of stable isotopes, I. Deuterium, 1971–1976, *Biomed. Mass Spectrom.*, **5**, 91–111; II. [13]C- 1971–1976, *Biomed. Mass Spectrom.*, **5**, 321–330; III. [15]N- 1971–1976, *Biomed. Mass Spectrom.*, **5**, 373–379; IV. [17]O, ^{18}O, [34]S, 1971–1976, *Biomed. Mass Spectrom.*, **5**, 425–432.

Klein, E. R., and Klein, P. D. (1979) 'A selected bibliography of biomedical and environmental applications of stable isotopes, V. [2]H, ^{13}C, ^{15}N, [18]O and [34]S, 1977–1978, *Biomed. Mass Spectrom.*, **6**, 515–545.

Klein, P. D., Hachey, D. L., Kreek, M. J., and Schoeller, D. A. (1978) 'Stable Isotopes: Essential tools in biological and medical research' in *Stable Isotopes: Applications in Pharmacology, toxicology and chemical research* (T. A. Baillee, Ed.), MacMillan, London, pp. 3–13.

Knapp, D. R., Gaffney, T. E., and McMahon, R. E. (1972) 'Use of stable isotope mixtures as a labelling technique in drug metabolism studies', *Biochem. Pharmacol.*, **21**, 425–429.

Ku, C. C., Kapoor, I. D., and Rosen, J. D. (1978) 'Metabolism of cytrolane (mephosfolan) systemic insecticide in a simulated rice paddy', *J. Agr. Food Chem.*, **26**, 1352–1357.

Ku, C. C., Kapoor, I. D., Rosen, J. D., and Stout, S. J. (1979) 'Photodegradation of cytrolane (mephosfolan) systemic insecticide in the aquatic environment using carbon-13 as a mass tracer', *J. Agr. Food Chem.*, **27**, 1046–1050.

Leete, E. (1977) 'The incorporation of [5,6-[13]C$_2$-] Nicotinic acid into the tobacco alkaloids examined by the use of [13]C- nuclear magnetic resonance', *Bioorganic Chemistry*, **6**, 273–286.

Lockhart, I. M. (1980) 'Stable isotopes—separation and application' in *Isotopes: Essential Chemistry and Applications* (J. A. Elvidge, and J. R. Jones, Eds.), Special publication No. 35, The Chemical Society, London, pp. 1–35.

Matthews, D. E., Ben-Galim, E., and Bier, D. M. (1979) 'Determination of stable isotopic enrichment in individual plasma amino acids by CI/ms', *Anal. Chem.*, **51**, 80–84.

Maynert, E. W. (1960) Metabolic fate of diphenylhydantoin in dog, rat and man, *J. Pharmacol. Exp. Ther.*, **130**, 275–284.

Maynert, E. W. (1965) 'The alcoholic metabolites of pentobarbital and amobarbital in man', *J. Pharmacol. Exp. Ther.*, **150**, 118–121.

Maynert, E. W., and van Dyke, H. B. (1950a) 'Metabolic fate of pentobarbital, isotope dilution experiments in urine after administration of labelled pentobarbital', *J. Pharmacol. Exp. Ther.*, **98**, 174–179.

Maynert, E. W., and van Dyke, H. B. (1950b) 'Metabolism of amytal labelled with [15]N- in dogs', *J. Pharmacol. Exp. Ther.*, **98**, 180–183.

Millard, B. J. (1978a) *Quantitative Mass Spectrometry*, Heydon and Son Ltd., London, pp. 50–51.

Millard, B. J. (1978b) 'Sources of error and criteria for the selection of internal standards for quantitative mass spectrometry' in *Quantitative Mass Spectrometry in the Life Sciences, II* (A. P. de Leenheer, R. R. Roncucci, and C. van Peteghem, Eds.), Elsevier, Amsterdam, pp. 83–102.

More, J. E., Roberts, T. R., and Wright, A. N. (1978) 'Studies of the metabolism of 3-phenoxybenzoic acid in plants'. *Pest. Biochem. Physiol.*, **9**, 268–280.

Nelson, S. D., Vaishnav, Y., Mitchell, J. R., Gillette, J. R., and Hinson, J. A. (1979) 'The use of ^{2}H- and ^{18}O- to examine arylating and alkylating pathways of phenacetin metabolism' in *Stable Isotopes: Proceedings of the Third International Conference* (E. R. Klein and P. D. Klein, Eds.), Academic Press, New York, pp. 385–392.

Ott, D. G. (1979) 'One-carbon ^{13}C-labelled synthetic intermediates. Comparison and evaluation of preparative methods', in *Stable Isotopes: Proceedings of the Third International Conference* (E. R. Klein and P. D. Klein, Eds.), Academic Press, New York, pp. 11–17.

Picart, D., Jacalot, F., Berthou, F., and Flock, H. H. (1978) 'Respective functions of carrier and internal standard in mass fragmentometric quantitative analysis' in *Quantitative Mass Spectrometry in the Life Sciences, II* (A. P. de Leenheer, R. R. Roncucci and C. van Peteghem, Eds.), Elsevier, Amsterdam, pp. 105–118.

Pohl, L. R., Nelson, S. D., Garland, W. A., and Trager, W. F. (1975) 'The rapid identification of a new metabolite of warfarin via a chemical ionization mass spectrometry ion doublet technique', *Biomed. Mass Spectrom.*, **2**, 23–30.

Rackham, D. M. (1980) '^{13}C-n.m.r. spectroscopy in medicinal chemistry' in *Isotopes Essential Chemistry and Applications* (J. A. Elvidge and J. R. Jones, Eds.), Special publication No. 35, The Chemical Society, London, pp. 97–122.

Rueppel, M. L., Brightwell, B. B., Schaefer, J., and Marvel, J. T. (1977) 'Metabolism and degradation of glyphosate in soil and water, *J. Agr. Food Chem.*, **25**, 517–528.

Rueppel, M. L., Malik, J. M., Brightwell, B. B., Marvel, J. T., Moran, S. J., Suba, L. A., and Nadeau, R. G. (1979) 'The use of ^{13}C-labelling in pesticide metabolite structure elucidation at agronomic use rates', *American Weed Science Society Conference*, **1979**, Abstract no. 20.

Rueppel, M. L., Marvel, J. T., and Schaefer, J. F. (1974). *The use of ^{13}C-labelled pesticides for metabolism studies in plants, animals and the environment*, presented at American Chemical Society Meeting, Sept, 8–13, Atlantic City, New Jersey, 1974.

Rueppel, M. L., Marvel, J. T., and Schaefer, J. (1975a) 'The spectral characterization of N-phosphonomethyl-glycine and its soil metabolites by ^{1}H, ^{31}P, ^{13}C-n.m.r. and gc/ms', division of pesticide chemistry Paper No. 25, in *Proceedings of the 170th National Meeting of the American Chemical Society*, Chicago, Illinois.

Rueppel, M. L., Marvel, J. T., Suba, L. A., and Schaefer, J. (1975b) 'The characterization of N-phosphonomethyl-glycine and its plant metabolites by n.m.r., derivatization, gc/ms/COM and isotopic dilution techniques', Division of Pesticide Chemistry Paper No. 27, in *Proceedings of the 170th National Meeting of the American Chemical Society*, Chicago, Illinois.

Shackleton, C. H. L., Hirota, H., and Honour, J. W. (1979) 'Application of deuterium labelling in mineralocoticosteroid analysis' in *Stable Isotopes: Proceedings of the Third International Conference* (E. R. Klein and P. D. Klein, Eds.), Academic Press, New York, pp. 37–45.

Still, G. G., Norris, F. A., and Iwan, J. (1976) 'Solubilization of bound residues from 3,4-dichloroaniline-^{14}C and propanil-phenyl-^{14}C treated rice root tissues' in *Bound and conjugated Pesticide Residues* (D. D. Kauffman, G. G. Still, G. D. Paulson, and S. K. Bandal, Eds.), ACS symposium series No. 29, ACS, Washington D.C., pp. 156–165.

Still, G. G., Balba, H. M., and Mansager, E. R. (1979) 'Characteristics and analysis of bound pesticide residues in plants', in *Abstracts from 4th International Congress of Pesticide Chemistry (IUPAC)*, July 24–28, 1978, Abstract V-26.

Tanaka, F. S., Swanson, H. R., and Frear, D. S. (1972) 'An unstable hydroxymethyl intermediate formed in the metabolism of 3-(4-chlorophenyl)-1-methyl urea in cotton', *Phytochemistry*, **11**, 2701–2708.

von Unruh, G. E., Hauber, D. J., Schoeller, D. A., and Hayes, J. M. (1974) 'Limits of detection of ^{13}C- labelled drugs and their metabolites in human urine', *Biomed. Mass Spectrom.*, **1**, 345–349.

Wall, M. E., Brine, D. R., Brine, G. A., Tomer, K., Davis, K. H. Jr., and Parker, C. J. (1979) 'Determination of paraquat in Marihuana pyrolysis products' in *Stable Isotopes: Proceedings of the Third International Conference* (E. R. Klein and P. D. Klein, Eds.), Academic Press, New York, pp. 129–137.

Weinkam, R. J., Gal, J., Callery, P., and Castagnoli, N., Jr. (1976) 'Application of chemical ionization mass spectrometry to the study of stereoselective *in vitro* metabolism of 1-(2,5-dimethoxy-4-methyl-phenyl)-2-aminopropane', *Anal. Chem.*, **48**, 203–209.

Wolen, R. L., Carmichael, R. H., Ziege, E. A., Quay, J. F., and Thompkins, L. (1979) 'Bioavailability studies utilizing stable isotope labelling—Benoxaprofen' in *Stable Isotopes: Proceedings of the Third International Conference* (E. R. Klein and P. D. Klein, Eds.), Academic Press, New York, pp. 475–483.

Progress in Pesticide Biochemistry, Volume 2
Edited by D. H. Hutson and T. R. Roberts
© 1982 John Wiley & Sons, Ltd

CHAPTER   3

Sugar conjugates of pesticides and their metabolites in plants—current status

V. T. Edwards, A. L. McMinn and A. N. Wright

INTRODUCTION

Early studies on xenobiotic metabolism often neglected conjugates or merely described them as hydrophilic or polar metabolites. One reason for this was the problem of isolating, and in particular characterizing, such hydrophilic substances. This was often coupled with the very small amounts that were available for examination—a problem which even in 1981 has still not been fully overcome. One noticeable early exception, was the work of L. P. Miller (1940) on the conjugation of o-chlorophenol in gladiolus corms. Using non-radiolabelled aglycones, Miller was able to isolate and characterize several glycosides and gentiobiosides. One feature of this and subsequent studies was the large amounts of conjugates available, up to 13% w/w in one case, which allowed isolation and quantification of crystalline acetates.

Since that period, striking developments have been made in analytical instrumentation, and many of the advances in chromatography and high-resolution spectroscopic techniques in principle simplify the task of isolating and characterizing pesticide metabolites. The inherent multidisciplinary nature of pesticide metabolism studies, encompassing chemistry, biochemistry and biology, has meant that the pesticide metabolism chemist or biochemist has been exposed to a changing environment. However, a fairly thorough review of the pertinent literature of pesticide conjugates in plants from 1975–1980 has revealed a paradox. With the abundance of powerful analytical methods now available, why have there been so few complete identifications of pesticide conjugates? It is still true that lack of availability of large amounts of conjugates, e.g. more than 1 mg, coupled with poor conjugate : coextractive ratios, markedly reduces the chances of a successful identification. In this context, a major factor affecting the outcome of a study can often be the procedures adopted for conjugate preparation, some of which will be discussed below. The application of tissue culture techniques to pesticide metabolism studies, especially for the preparation of conjugates, results in a relatively high conjugate : coextractive ratio.

Also, in recent years our understanding of the physiological significance of conjugates in plant systems has altered. Pesticide conjugates, like naturally occurring secondary metabolites (Bu'lock, 1964) have often been considered to have no obvious metabolic function. The concept of a dynamic 'metabolic pool', where conjugates are temporary, as well as permanent, storage points, is perhaps a more accurate view nowadays.

Nonetheless, advances will be made in our understanding of the biochemical role of pesticide conjugates, only when systematic progress is achieved in the identification of pesticide conjugates.

This chapter provides an attempt to outline current trends in the study of pesticide sugar conjugates formed by plants, and will attempt to summarize current knowledge on the chemistry and biochemistry of plant sugar conjugates.

Aspects of the subject have been reviewed by several authors, the most recent comprehensive reviews being those by Frear (1976) and Dorough (1979).

TECHNIQUES FOR PREPARATION, ISOLATION, CHARACTERIZATION AND DETERMINATION OF PESTICIDE SUGAR CONJUGATES

Preparation of conjugates

The various approaches to preparation of pesticide conjugates can be readily classified on the basis of conjugate : coextractive ratio e.g. from normal field application (low ratio) to tissue culture (high ratio). The choice of preparative method is one of the most significant factors in a successful conjugate study.

Normal field application

Since the conjugates obtained may well vary with the preparation conditions, the only genuinely valid way of preparing them for study will be under normal, recommended use. However, this precludes the use of radiolabelled compounds The optimum is to use radiolabelled compounds with procedures approaching normal use as closely as possible, e.g. in outdoor enclosures. Conjugates identified from such experiments should be representative of those likely to occur in practice, but amounts available for identification from outdoor 'realistic' experiments are inevitably very small. However, if a residue analysis method for conjugated metabolites is developed, this can be used to monitor field samples. These procedures are discussed in a subsequent section.

Outdoor metabolism studies

In practice, such studies, using realistic dosage rates of radiolabelled pesticide under normal crop growth conditions, should produce metabolites qualitatively and quantitatively approaching those formed under normal agricultural practice, but with the advantage of radiolabel. Such studies can be most valuable in understanding the actual mechanism of pesticide metabolism. The primary disadvantage, as pointed out above, is that the concentrations of pesticide conjugates formed will be low, and the likelihood of complete identification of conjugate is often reduced. For example, in a study of the metabolism of [14C]Aldicarb ([2-methyl-2-(methylthio)propionaldehyde-o-methylcarbamoyloxime]) in sugar beet (*Monyx monogerm* var.), (Rouchaud et al., 1980) it was found that 54 % of the total 14C-radioactivity incorporated into the whole leaves was water-soluble, which suggested that it was in the form of plant conjugates, although no identification was achieved. Using a commercially available formulation, the sugar beet plants were treated in-furrow at sowing at a dosage rate three times higher than recommended.

Thus it has been usual for workers to report the aglycones formed on hydrolysis, often using a fairly non-selective enzyme preparation or other reagent (Gorbach et al., 1977; Bull et al., 1976). Since such studies tend to form the basis of most work on pesticide degradation, they have, in part, contributed to the general view that identification of conjugates is very demanding.

An extensive outdoor study of the conjugate formation of the insecticide [^{14}C]oxamyl in a variety of plant tissues has been carried out (Harvey et al., 1978). In this study, potatoes, alfalfa, peanuts, apples and oranges were grown and treated under typical field conditions. It was possible to show that the predominant polar species of short-term studies was a glucoside of the non-toxic oximino metabolite (1)

$$
\begin{array}{c}
CH_3 \\
\diagdown \\
\text{N—C—C=N—O-glucose} \\
CH_3 \quad\; \overset{\displaystyle O}{\|} \quad\; | \\
SCH_3
\end{array}
$$

(1)

and at medium term the monomethyl oximino metabolite (2) was also formed.

$$
\begin{array}{c}
CH_3 \\
\diagdown \\
\text{N—C—C=N—O-glucose} \\
H \quad \overset{\displaystyle O}{\|} \quad | \\
SCH_3
\end{array}
$$

(2)

Both species apparently underwent further conjugation with additional hexose units to the polymer (3).

$$
\begin{array}{c}
CH_3 \\
\diagdown \\
\text{N—C—C=N—O-glucose-(hexose)}_n \\
R \quad \overset{\displaystyle O}{\|} \quad | \qquad\qquad R = H, CH_3 \\
SCH_3
\end{array}
$$

(3)

In some instances, outdoor metabolism studies using realistic application rates of pesticide have been complemented by parallel glasshouse studies running simultaneously, e.g. cis- and trans-permethrin (Gaughan and Casida, 1978) and decamethrin (Ruzo and Casida, 1979).

Using a combination of laboratory and field grown wheat and soil, the degradation of the urea herbicide, chlortoluron (4) was studied by Gross et al. (1979). In the young wheat plants, the main degradative mechanism was the oxidation of the ring methyl to form benzylalcohol derivatives (5) which were readily conjugated. Evidence for the view that conjugates are not stable metabolites was provided by the observation that in the older plants these benzylalcohol conjugates were cleaved before the free benzylalcohol derivatives could be oxidized to the corresponding benzoic acid derivatives (6).

$R_1 = H$ or CH_3
$R_2 = H$ or CH_3

Undoubtedly the extent of such studies will depend on the radiochemical regulations of the country in which they are conducted. With long-lived isotopes, e.g. ^{14}C, it is probably advisable to conduct studies under controlled conditions in such a way that the experiment can be subsequently disposed off. Such an approach was used in a study of the metabolism of the pyrethroid insecticide, cypermethrin, in lettuce, using a custom-built outdoor radiochemical enclosure (Wright et al., 1980). In this wire covered enclosure, lettuce plants were grown in a wooden box filled with fresh field soil, standing in water in an aluminium tray to provide bottom watering. In this way complete control of radioactivity, other than through volatilization, was achieved.

Laboratory metabolism studies using intact plants

An intermediate type of procedure between the use of plants in soil and the more artificial laboratory preparations such as excised leaves, has been the use of intact plants maintained in nutrient solution in the absence of soil. One advantage of this approach is that extremely good root uptake of nutrients and radio-labelled pesticide can be achieved while maintaining the integrity of a complete plant system. However, maintenance of the plants involves more work than in conventional soil–plant studies, and due care and attention should be paid to maintaining a correct nutrient balance. Kunstman and Lichtenstein (1979) showed that corn plants (Funk hybrid 9444) grown in Hoagland's nutrient solution incubated with $[^{14}C]$diazinon, displayed a reduced metabolism of the pesticide when the medium was calcium- or nitrogen-deficient. Thus, roots contained increased amounts of organo-soluble metabolites and decreased levels of water-soluble metabolites.

An extensive study of the metabolism of the soil fungicide, 3-hydroxy-5-methylisoxazole, in cucumber (*Cucumis sativus* L., 'Midorifushinari') seedlings and excised root tips was carried out in hydroponic solution (Kamimura *et al.*, 1974). With plants maintained in a glasshouse under a programmed lighting/ heating regime, these authors isolated and identified two isomeric glucose conjugates, 3-(β-D-glucopyranosyloxy)-5-methylisoxazole (**7**) and 2-(β-D-gluco-pyranosyloxy)-5-methyl-4-isoxazoline-3-one (**8**). The rates of formation and the relative proportions of (**7**) and (**8**) were extremely dependent on a variety of environmental and morphological factors, and both compounds were shown to exhibit different biological properties.

(7)   (8)

Rice plants (*Oryza sativa* L. var. Nato) grown in full strength Hoagland's solution and exposed to propanil (3,4-dichloropropionaniline) were shown to form three conjugated metabolites of 3,4-dichloroaniline (Still, 1968). Only one was characterized as *N*-(3,4-dichlorophenyl)-glucosylamine. A time-course study revealed that this conjugate reached a maximum concentration after four days and decreased thereafter.

To provide a protective matrix for the exposed roots of 3–4 year old mugho pines (*Pinus mugo* Turra), a bed of glass spheres was used, which was surrounded by the nutrient solution containing [^{14}C]carbofuran, (Pree and Saunders, 1974). As this study progressed the concentration of metabolites in the aqueous phase increased, with most of this activity being hydrolysed by β-glucuronidase or β-glucosidase.

Other recent studies in which intact seedlings or plants have been grown in nutrient medium, and in which conjugated metabolites have been found and identified include those by Still and Mansager (1972) (isopropyl 3-chlorocar-banilate), Hodgson and Hoffer (1977a,b) (diphenamid), Shimotori and Kuwat-suka (1978) (2,4,6-trichlorophenyl-4'-nitrophenyl ether) and Donald and Shimabukuro (1980) (diclofop-methyl). In addition similar studies in which conjugated metabolites have been proposed but have not been identified include Starr and Cunningham (1975) (4-aminopyridine), Jordan *et al.* (1975) (siduron), Hill and Krieger (1975) (tirpate), Niki *et al.* (1976) (chlomethoxynil), Kohli *et al.* (1976) (lindane), Ogawa *et al.* (1976) (BPMC), Motooka *et al.* (1977) (Sandoz 6706), Mangeot *et al.* (1979) (metribuzin), Domir (1978) (maleic hydrazide).

It would appear that in recent years, the trend is away from carrying out metabolism studies on intact plants and conventional soil-based media, towards a more general use of nutrient media under controlled environmental conditions.

Plant preparations under laboratory conditions

For preparation of larger amounts of pesticide conjugates it is possible to treat individual plant sections under laboratory conditions. Abscised leaves and roots, leaf discs, seedlings and shoots are typical plant sections and uptake of the labelled pesticide is normally fairly rapid from aqueous nutrient solution. Some useful applications of this type of preparation will be discussed followed by a brief discussion of the potential of such experimental plant systems in conjugate studies.

Frear et al. (1978) germinated seeds of giant foxtail (Setaria sp.) and barley (Hordeum vulgare L.) on moist paper tissue at 25 °C in the absence of light. Tissue sections of roots, shoots and hypocotyls were taken from 3–4 day old seedlings and after distribution on the bottom of beakers, were treated for varying periods (up to 24 h) with a solution of [^{14}C]chloramben. In addition to uptake studies of the pesticide, they also studied the role of isolated [^{14}C]-chloramben glucose ester and [^{14}C]-N-glucoside metabolites in susceptible (barley) and non-susceptible (soya bean) species, using excised tissues and seedlings. The significance of their findings is discussed in the 'Interconversion reactions of pesticide conjugates' section (p. 110).

The use of abscised leaf sections has been extremely valuable in a series of studies on pyrethroid metabolism in plants. In studies of the metabolism of 3-phenoxybenzoic acid (More et al., 1978) and trans-2(2',2'-dichlorovinyl)-3,3-dimethylcyclopropane carboxylic acid (Wright et al., 1980), the authors used the following technique on a range of commercial crops. Abscission of leaves and roots was carried out while the young plants were immersed in water, thus ensuring that an embolism due to a sudden uptake of air at the freshly cut surface did not occur. The cut section was then transferred, while moist, to a vial containing an aqueous solution of the radiolabelled pesticide. Uptake of the pesticide was found to be rapid and the vials were replenished with distilled water as required. Plants were extracted and analysed at up to 72 h after treatment.

A comparison of the absorption of [^{14}C]barban by intact and sectioned shoots of wheat (cv. Ushio) and oat (cv. Victoria) has been reported (Kobayashi and Ishizuka, 1977). Shoots of intact plants were dipped in an aqueous solution containing 1.5 mg. kg^{-1} [^{14}C]barban. Shoot sections (about 1 cm) were incubated at 30 °C in 0.01 M potassium phosphate buffer containing 1.5 mg. kg^{-1} [^{14}C]barban for periods up to 9 h. These authors found that the rates of absorption by sectioned shoots were almost identical in the susceptible (oat) and non-susceptible (wheat) species although the rates differed markedly from those by the intact roots. It was suggested that the amount of absorption from the intact root surface is small compared with the rapid uptake at the abscissions of the shoot sections.

Kamimura *et al.* (1974), in a study of the uptake of the soil fungicide F-319, suspended cucumber (*Cucumis sativus* L.) seedlings at the two-leaf stage in air after removal from nutrient solution. While suspended, the rootlets were separated individually and after covering the remaining exposed parts of the roots with moist cotton wool, the rootlets were immersed in a nutrient solution containing [^{14}C] F-319. After a 24 h period, the plants were dried and autoradiographed. Uptake of the [^{14}C] F-319 was found to be fairly rapid and general throughout the root system, although a major disadvantage of the technique, which precluded a quantitative assessment of the metabolism, was the diffusion of ^{14}C into the surrounding absorbent cotton wool.

The formation of glycoside conjugates of decamethrin and its main metabolites in leaf discs of cotton (*cv.* Stoneville FA) and snapbeans (*cv.* Contender) has been studied (Ruzo and Casida, 1979). Discs (10 mm in diameter) were punched out with a cork borer from cotton and bean plants under water. The discs were incubated at 30 °C for 5 h in an aqueous solution (2 ml) containing 1–5 μg of ^{14}C substrates.

The metabolism of [1,2-^{3}H] gibberellin A$_1$ (GA$_1$) in excised lettuce (*Lactuca sativa* L.) hypocotyl has been studied by Stoddart and Jones (1977). Using the method of Silk and Jones (1975) uptake of [^{3}H]GA$_1$ was shown to be linear for 24 h and that formation of polar, water soluble metabolites occurred over this period. Using a partitioning technique, acrylamide gel-filtration and electrophoresis, Stoddart and Jones were able to provisionally identify GA$_1$-glucosyl ether (**9**) and GA$_1$-glucosyl ester (**10**).

The value of the techniques described above is that they are relatively easy to carry out and that significant amounts of conjugates can be isolated. Clean-up of the initial plant extract is often simplified, since the proportion of coextractives should be reduced compared with whole plant studies. Carefully executed, such methods provide the metabolism chemist with a very useful tool. However, the data obtained must be seen in the context of the inherent limitations of the approach. A major question is whether the conjugates formed at high concentrations are the same as those formed under field conditions. The abnormal dosage of aglycone may well inhibit the formation of conjugates of a minor group, so that atypical formation of a glucose or amino acid conjugate predominates. Essentially the conjugates formed in abscised tissue studies reflect only the metabolism over a limited period, usually less than 24 h, and correlation of the metabolism with longer term studies using intact plants should be treated with caution. Nonetheless, providing these limitations are accepted, such techniques provide a fairly rapid and precise means for the isolation of larger amounts of 'clean' conjugate.

Cell culture techniques

Apart from purified enzyme preparations, the simplest system available for metabolism studies is cultured plant cells. These came into active use about 1970 (Locke *et al.*, 1971; Feung *et al.*, 1971) and are now in use in many laboratories since the procedures are relatively easy to set up and the cultures often give ready metabolism of pesticides. One unique advantage is good contact between the pesticide and the cells with avoidance of the penetration delays associated with whole plants. The culture solutions are relatively clean, easy to analyse and can be treated with high concentrations of pesticide. Water-insoluble pesticides can be added in small volumes of solvent (e.g. dimethylsulphoxide) or detergent. For example, benzo[α]pyrene was dispersed in lecithin for the application (Trenck and Sandermann, 1978).

However, there are limitations since although cultured cells are genetically complete, parts of the metabolic processes, such as photosynthesis, are 'switched-off'. Thus conversions detected are likely to be realistic but may only represent part of the changes that would occur in intact plants. Moreover, products which are normally minor intermediates may build up. The rapid metabolism often found is balanced by normal use of rather limited (up to 14 day) periods of metabolism.

The use of cell culture techniques in xenobiotic metabolism studies has recently been reviewed (Sandermann *et al.*, 1977; Mumma and Hamilton, 1979) and the reviews contain many useful examples of studies. Indeed, a variety of pesticides have been applied to cell cultures ranging from non-polar compounds such as DDT and kelthane (Scheel and Sandermann, 1977) and aldrin (Brain and Lines, 1977) to more polar ones such as carbaryl (Locke *et al.*, 1976), cisanilide (Frear and Swanson, 1975), chlorpropham (Davis *et al.*, 1977), diphenamid (Davis *et al.*, 1978) and diclofop-methyl (Dusky *et al.*, 1980). Mumma's group in Pennsylvania published an important series of papers on 2,4-D and related compounds (Feung *et al.*, 1978 and others). These papers and that of Frear and Swanson (1975) consider the relation of such studies to results in whole plants and conclude that substantially the same pathways exist but that the proportions vary. In particular there is increasing evidence that cell cultures do not form glycosides as readily as whole plants. In this context, the study with IAA in soyabean root callus culture (Davidonis *et al.*, 1980) might be compared with studies by Zenk (1963) and Bandurski (1979) (considered on pp.114–115). Davidonis *et al.* (1980) also found that interconversions occurred, the IAA apparently being converted into conjugate and then being released again. Such variation could result from varying nutrient concentrations in the culture, but also to two or more conjugates being formed in proportions varying with time. Thus the larger amino acid conjugate yields often found in cultures may be due to initial formation of sugar conjugates which are not easily detected owing to subsequent ready conversion into amino acid conjugates.

Another variable can be the cultures themselves. Feung *et al.* (1975) applied 2,4-D to five species of plant culture for comparison and found that most hydroxylation occurred in maize culture, aspartic acid conjugates were found in maize, tobacco and jackbean, while all these, plus carrot and sunflower, formed glutamic acid conjugates. Glucosides were also probably formed in all the species. Another factor, less understood, may be the physical state of the culture. Normally, callus cultures are maintained as agglomerations of cells growing on a gelled nutrient. These can be treated directly with pesticide to the nutrient, or as a pool on the surface of the gel. More commonly, the pesticide has been dispersed in liquid nutrient and the lumps of callus were slurried in the gently swirled nutrient. Finally, dispersed suspension cultures can be prepared and maintained. Undoubtedly the second procedure is the simplest to carry out (e.g. with 10 g cells to 25–100 ml nutrient) and appears to be most often used.

Finally, it must be said that the results obtained can vary with the treatment conditions, e.g. age of callus, although it is likely that in many cases the qualitative results will be the same. One advantage of callus cultures may be a greater production of glycosides of compounds retained in the callus, while suspensions of cells tend to release the free metabolic products back into the culture medium. An extension of this feature could be variable results where a suspension culture contains some aggregates of adhering cell masses.

Isolation of intact conjugates

For convenience, the isolation and purification of intact conjugates can be considered in three phases: extraction, primary clean up (e.g. solvent partiton) and secondary clean up (usually involving some form of chromatography).

Table 1 provides some data on typical isolation procedures in the recent literature.

Extraction

The relatively high polarity of pesticide glycoside conjugates requires the use of a polar solvent for extraction from a plant matrix. However, in many pesticide metabolism studies, using radiolabel, the aim is to extract the maximum amount of labelled metabolites which will often comprise a fairly broad range of polarities. For reasonably efficient extraction of glycoside conjugates, aqueous mixtures of acetone, acetonitrile or alcohols should be sufficient. For conjugate isolation it is ultimately the aqueous extract that is required, and the organic cosolvent should be relatively easily removed e.g. by rotary evaporation. This is particularly important for labile components e.g. the malonylglucosyl ester of flampropmethyl (Dutton *et al.*, 1976). Too few genuine extractability studies for specific pesticide conjugates have been carried out (see conjugate analysis section) although if radiolabel is used in the study, an in-built mechanism exists for over-all extractability. In fact the main potential problem associated with the choice of organic solvents is the formation of artefacts. Saleh *et al.*, (1980) recently pointed out potential problems in storing α-cyanophenoxybenzyl pyrethroids in acetone due to the formation of 2-(3-phenoxybenzoyl)-2-propyl esters.

Methanol is widely used as an extracting solvent either alone, in aqueous solution or in combination with chloroform. Prolonged exposure of conjugates to methanol is not advisable due to the possibility of trans-esterification of ester-linked conjugates. While free carboxylic acid groups do not usually react unless mineral acid is present, the presence of plant enzyme systems can induce fairly rapid and complete trans-esterification (Pillmoor *et al.*, 1981). In related studies on abscisic acid, Milborrow (1970) used methanol to extract plant tissue, but added BHT as an antioxidant. In a subsequent study (Milborrow and Mallaby, 1975) the formation of methyl abscisate was shown to be due directly to the choice of extracting solvent. The use of dry methanol as an extracting solvent obviously depends on the plant moisture content to increase extractability. However, where disaccharides or more complex conjugates are suspected, which would be poorly extracted in such a system, the use of at least 20 % water is recommended.

The formation of ethyl esters as artefacts has been demonstrated in a tissue culture study of 2,4-D glucoside formation (Feung *et al.*, 1976). In an earlier study, Zenk (1961) had shown that the glucose ester of IAA was unstable on t.l.c. in the presence of solvents containing ammonia.

A well-established procedure, generally known as the Bligh–Dyer procedure, uses a mixture of chloroform, methanol and water. It is an established method for the extraction of lipophilic substances (Haque *et al.*, 1978) but poses a

Table 1 Isolation and clean-up of pesticide conjugates

Pesticide	Crop	Conjugate	Extraction solvent (v/v)	Primary clean-up	Secondary clean-up of aqueous phase	Reference
Chloramben	foxtail barley	α-D-glucoside	methanol		1. Amberlite XAD-2 2. t.l.c. 3. Biogel P2	Frear et al. (1978)
Chlortoluron	wheat	glucosides	methanol–water (80:20)	hexane, diethyl ether–water partitions	1. Membrane ultrafiltration 2. Sephadex G15	Gross et al. (1979)
Cypermethrin	lettuce	β-D-glucoside	1. acetone 2. acetonitrile–water (70:30)	petroleum spirit, ethyl acetate–water partitions	1. Acetylation 2. t.l.c. 3. h.p.l.c.	Wright et al. (1980)
Diclofop-methyl	wheat wild oat	glucosides	methanol–water (80:20)	diethyl ether–water partition	1. Amberlite XAD-2 2. Sephadex LH-20 3. t.l.c.	Shimabakuro et al. (1979)
Diphenamid	peppers	glucosides	chloroform–water	1. Amberlite XAD-2 2. DEAE cellulose 3. Biogel P2	1. Sephadex LH-20 2. Poropak Q	Hodgson and Hoffer (1977b)

Compound	Source	Conjugate	Extraction/Partition	Purification	Reference
Flamprop	wheat	malonyl glucosides	acetone–water (1:1) ethyl acetate–water partition (pH 4.0)	1. t.l.c. (Cellulose F) 2. h.p.l.c. 3. Electrophoresis	Dutton *et al.* (1976)
3-Hydroxy-5-methylisoxazole	cucumber	glucosides	ethanol–water (90:10) 1. n-hexane, chloroform–water partitions 2. dialysis of aqueous phase	1. Sephadex LH-20 2. Silica gel column chromatography	Kamimura *et al.* (1974)
Isopropyl-3-chloro-carbanilate	soya bean	*O*-glucoside	1. methanol–chloroform–water (2:1:0.3) (2:2:1.8) 2. methanol–water n-butanol–water partition	1. Aluminium oxide column chromatography 2. butan-1-ol extraction 3. DEAE cellulose column	Still and Mansager (1972)
Oxamyl	various	glucose conjugates	methanol hexane–water partition	h.p.l.c. on 1. Biolgel P2 2. Sephadex LH-20 3. Porasil A	Harvey *et al.* (1978)
Perfluidone	peanuts	β-*O*-D-glucosides	methanol–water (80:20) chloroform–water partition	1. Amberlite XAD-2 2. Biogel P2 3. t.l.c. 4. h.p.l.c.	Lamoureux and Stafford (1977)

problem over control of the crop moisture : solvent ratio, in addition to using a toxic solvent.

Primary and secondary clean-up

Having obtained a crude plant extract, the primary step in the process of obtaining conjugate in a reasonably pure form usually involves solvent partition. After removal of the excess organic solvent from the plant extract, the crude aqueous residue remaining is generally partitioned between an organic solvent and water. Hydrocarbon solvents, e.g. methylene dichloride, do not extract polar conjugates whose polarity is largely determined by the sugar moiety. Oxygenated solvents, e.g. diethyl ether or ethyl acetate, are of more value for extracting moderately polar conjugates, e.g. monosaccharide conjugates. However, for disaccharides and more complex polar conjugates butan-1-ol will be required.

In many instances, the removal of plant coextractives is incomplete after the primary clean up stage and an extensive secondary clean-up of the aqueous extract is required (Table 1). Many of these methods are based on chromatography, and as well as affording clean up of the extract, some indication is usually provided, for the first time in the process, of the type of conjugate present. However, the full use of chromatographic methods in the characterization of conjugates will be dealt with in the next section.

Characterization of pesticide sugar conjugates

Having isolated and purified a sufficient quantity of the intact conjugate from a crude extract, in many cases having to extract the substance from a fairly complex biological matrix, the question now has to be asked: 'How do we know we have a conjugate?' This is probably best discussed in the context of the types of method chosen to characterise the suspected conjugates. These can be conveniently separated into three classes:

1. Chromatographic techniques
2. Spectroscopic techniques
3. Chemical and enzymatic techniques

Chromatographic techniques

Changes in chromatographic mobility with variations in solvent polarity can furnish very useful gross structural information on the conjugate. This applies equally to t.l.c., h.p.l.c. and other forms of liquid chromatography, e.g. gel permeation chromatography. Additionally, changes in chromatographic behaviour before and after hydrolysis or chemical derivatization, e.g. silylation, are of great value. Such chromatographic methods are essentially non-destructive, although losses, particularly of highly polar compounds, from chromatographic plates or columns, can be unacceptably high. However, derivatization combined with chromatography, as well as providing useful structural details, can also afford additional clean-up. In most cases, this balance between increased information and loss of sample will be decided by the quantity of conjugate to be handled.

Thin-layer chromatography (t.l.c.)   Undoubtedly, t.l.c. of conjugates either before or after chemical derivatization has provided one of the main methods by which their presence has been established. An excellent review by Heirwegh and Compernolle (1979) on the micro-analytical detection and structure elucidation of ester glycosides details a series of simple chemical treatments and t.l.c. separations which can not only reveal structural information but often confirm postulated structures. Using the minimum of standard laboratory equipment, chemical derivatization of glycosides, hydrolysis with glycosidases, methanolysis of glycosides, procedures for establishing the ester linkage and determination of sugar and aglycone after application of glycosides to a t.l.c. plate are discussed. The scope of this chapter prohibits further discussion in this section beyond citation of several recent references in which t.l.c. (often as two-dimensional cochromatography with authentic standards) in conjunction with enzyme or chemical hydrolysis has been used as the primary means of establishing the presence of conjugates: Gaughan and Casida (1978), Domir (1978), Ogawa *et al.* (1976), Marquis *et al.* (1979), Umetsu *et al.* (1979) and Ruzo and Casida (1979).

Since the majority of pesticide conjugate studies are carried out with radio-labelled pesticides, the use of radio-t.l.c. is of primary importance. For a review on developments and applications of radio-t.l.c. the reader is referred to a recent monograph (Roberts, 1978), although since its publication further advances in sensitivity have been achieved with the development of a linear analyser, which should increase the sensitivity of radio-t.l.c. by a factor of 10–50.

High-performance liquid chromatography (h.p.l.c.)   In recent years h.p.l.c. has developed rapidly although the advances in column design and packing materials have not really been balanced by equivalent developments in the range of detectors available. As a technique for clean-up of radiolabelled aqueous extracts, h.p.l.c. in its reverse phase mode has undoubted advantages to offer and the application of h.p.l.c. to pesticide metabolism studies has recently been reviewed (Harvey, 1980). As a technique for characterizing glycoside conjugates, radio-h.p.l.c. has not yet been fully exploited, despite its potential. However, there are examples of its use in the literature. For example, in a study of the conjugation of cypermethrin metabolites in lettuce, Wright *et al.* (1980) used radio-h.p.l.c. on a 5 μm Hypersil ODS column with acetonitrile : water (27 : 75, v/v) as mobile phase to confirm the identity of the glucose ester of the cyclopropane carboxylic acid (11).

$$
\begin{array}{c}
\text{H} \quad \text{CH}{=}\text{CCl}_2 \\
\diagup \hspace{-0.3em} \diagdown \\
\text{CH}_3 \hspace{-0.5em} -\hspace{-0.8em}\bigtriangleup\hspace{-0.8em}- \hspace{-0.5em}\text{COOR} \\
\text{CH}_3 \qquad \text{H}
\end{array}
$$

(11) R = glucose

Thin-layer cochromatography with reference compounds before and after hydrolysis, methylation and acetylation combined with chemical ionization g.c./m.s. of the methylated aglycone were used as confirmation of structure.

Harvey *et al.* (1978) made notable use of h.p.l.c. not only for extensive clean-up of metabolic extracts with a variety of column packings (Bio-Gel P-2, Sephadex LH-20 and Porasil A) but also in conjunction with radio-g.l.c. and g.c./m.s., to characterize two glucose conjugates of the insecticide/nematicide, oxamyl.

Gas-liquid chromatography (g.l.c.) Since its inception (James and Martin, 1952) there have been relatively few major changes in the basic technique, although in recent years the design of novel detector systems, the advent of capillary columns in a variety of forms and the synthesis of highly polar stationary phases have assisted in its development. There have also been relatively few major changes in radio-g.l.c., which is of more value in metabolism studies with radiolabel, and only recently the first report of successful radio-capillary g.l.c. was published (Gross *et al.*, 1980). Nonetheless, both g.l.c. and radio-g.l.c. can be of value in identifying structures of conjugates providing suitable derivatives can be prepared, and that standard reference compounds are available for cochromatography.

In recent years, the emphasis has been on the successful combination of the separating powers of g.l.c. and the analytical sensitivity of mass spectrometry, in g.c./m.s. Appropriate examples of conjugate identification by g.c./m.s. will be dealt with in the following section on mass spectrometry. However, one recent development has been the application of g.l.c. to separate optical isomers of pesticides and metabolites. The use of chiral stationary phases and adsorbents has been the subject of a recent review (Blaschke, 1980) although a fairly recent publication (Sakata and Koshimizu, 1979) demonstrated the optical resolution of the acetyl glucoside derivatives of DL-menthol and DL-borneol on g.l.c. with NPGS and SE stationary phases.

Spectroscopic techniques

These are specific instrumental methods to detect the presence of, and the surrounding chemical environment of, specific functional groups. Undoubtedly the potentially most powerful technique is mass spectrometry either alone in one of its increasingly numerous forms, e.g. chemical ionization (CI/m.s.), electron impact (EI/m.s.) or field desorption (FD/m.s.), or linked to gas chromatography (g.c./m.s.) and liquid chromatography (l.c./m.s.). However, other spectroscopic techniques, in particular nuclear magnetic resonance (n.m.r.) spectroscopy, and less so infrared (i.r.) and ultraviolet (u.v.), can also be of considerable value, providing that sufficient sample is available.

Mass spectrometry (*m.s.*)   The developments in mass spectrometry over the past decade have undoubtedly placed the technique at the forefront of analytical research, as the most powerful, and potentially most informative, method available for the structure elucidation of pesticide conjugates. The widespread use of the technique and its general acceptance have been reflected in a series of useful review articles in recent years which outline the potential and value of m.s. to the chemistry of natural products and biological molecules.

Of general interest are articles by Beckey and Schulten (1975) on field desorption m.s. (FD/m.s.), Morris (1980) on biological structure determination by m.s., and McLafferty (1980) on tandem mass spectrometry (m.s./m.s.). Of more specific interest are reviews on g.c./m.s. of plant glycosides (Martinelli, 1980), m.s. and g.c./m.s. of intact glucuronides (Fenselau and Johnson, 1980), and FD/m.s. of cyanogenic glycosides (Dreifuss *et al.*, 1980).

The use of FD/m.s. was of value in identifying the structure of the malonyl-glucosyl ester of flamprop-methyl (see below) a wild oat herbicide and the spectrum is illustrated in Figure 1.

Other recent examples where m.s. or g.c./m.s. have played a primary role in identification of pesticide glucoside conjugates and/or their aglycones include:

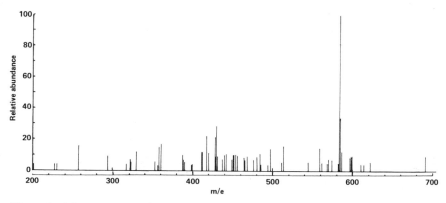

Figure 1   Mass spectrum of malonyl glucosyl ester of flamprop-methyl after methylation, determined on a Varian MAT-311A mass spectrometer with a field desorption source. Reproduced from *Chemosphere*, **3**, 200 (1976) with permission

DIB (Gross and Schütte, 1980), oxamyl (Harvey *et al.*, 1978), isopropyl-5-chloro-2-hydroxycarbanilate (Still and Mansager, 1972), diclofop-methyl (Shimabukuro *et al.*, 1979) and chlortoluron (Gross *et al.*, 1979).

In recent years l.c./m.s. emerged as potentially an ideal analytical tool for conjugate identification, although problems still exist with the l.c./m.s. interface (Games *et al.*, 1981). Nonetheless, the ability of h.p.l.c., in particular using reversed phase columns with polar solvent systems, to separate and clean-up sugar conjugates, coupled with the identification capabilities of m.s., is probably the most potentially useful technique which will be available to the pesticide chemist. In particular, the use of m.s. in the field desorption mode, should avoid the necessity to derivatize the conjugate prior to analysis.

Nuclear magnetic resonance spectroscopy (n.m.r.)   The correlation of conjugate structure and n.m.r. spectra is a complex task, but the assignment of structure to spectral characteristics has been simplified in recent years with the development of high resolution spectrometers. Nonetheless, the excessive costs of such instrumentation will ensure that only the larger research establishments will have access unless a contract system can be developed. Several interesting reports have appeared in recent years highlighting the value of n.m.r. on both high and low-field instruments, for conjugate identification. More *et al.* (1978) used ^{1}H-n.m.r. to assist in the structure identification of the glucosyl ester of 3-phenoxybenzoic acid (Figure 2). Frear *et al.*, (1978) made extensive use of ^{1}H-n.m.r. chemical shifts and coupling constants to show that a water-soluble metabolite of chloramben was present as the α-D-glucosyl ester, rather than the β-D-glucosyl ester previously reported (Frear *et al.*, 1977). The downfield shift and the small coupling constant of the anomeric proton doublet signal from the acetylated glucosyl ester was used to characterize the α-(**12**) rather than the usual β-configuration.

(12)

The value of ^{13}C-n.m.r. in distinguishing isomers in compounds which contain more than one centre of asymmetry has been clearly illustrated by Cairns *et al.*, (1978). In a study of the analytical chemistry of amygdalin (**13**) the authors were able to distinguish between the stable chiral centres of the gentobiose moiety and the unstable chiral centre of the aglycone, mandelonitrile. By careful assignment of the resonances, they were able to establish the optical and chemical purity of a range of amygdalin extracts. The application of ^{13}C-n.m.r. for the elucidation of the structures of natural products, e.g. flavonoid *C*-glucosides

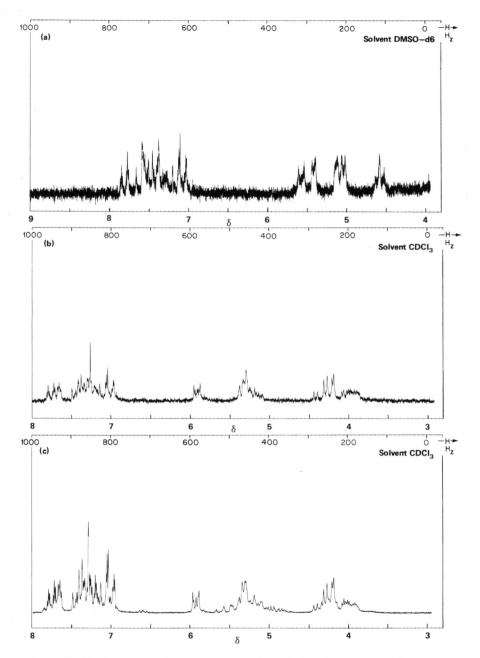

Figure 2 Nuclear magnetic resonance spectra of β-D-glucopyranosyl-3-phenoxybenzoate run (a) after partial purification, (b) after acetylation and repurification. Spectrum (c) shows the spectrum of a synthetic sample of 2,3,4,6-tetraacetyl-β-D-glucopyranosyl-3-phenoxybenzoate. Reproduced from *Pesticide Biochemistry and Physiology*, **9**, 273 (1978) by permission of Academic Press Inc.

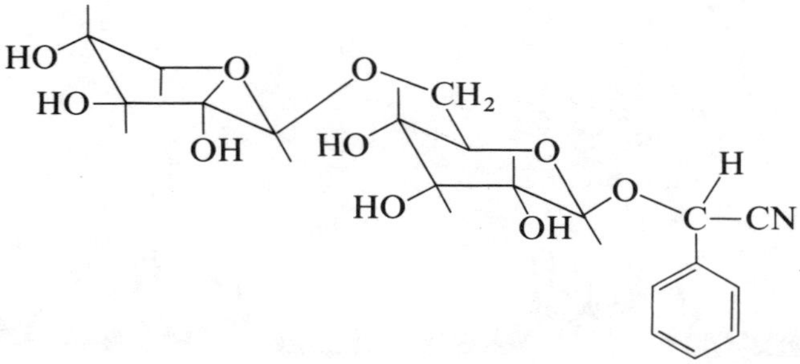

(13)

(Österdahl, 1978) and phenolic glycosides (Ozawa and Takino, 1979) provides an excellent source of ^{13}C-n.m.r. shifts which could be valuable for the interpretation of ^{13}C-n.m.r. spectra of pesticide glycoside conjugates.

In a detailed study of the acidic conjugate of the wild oat herbicide, flamprop-methyl, ^{19}F-n.m.r. indicated the presence of a single fluorine-containing component at 236.3 Hz, which was tentatively identified as the following malonyl-glucosyl ester (14) (Dutton *et al.*, 1976).

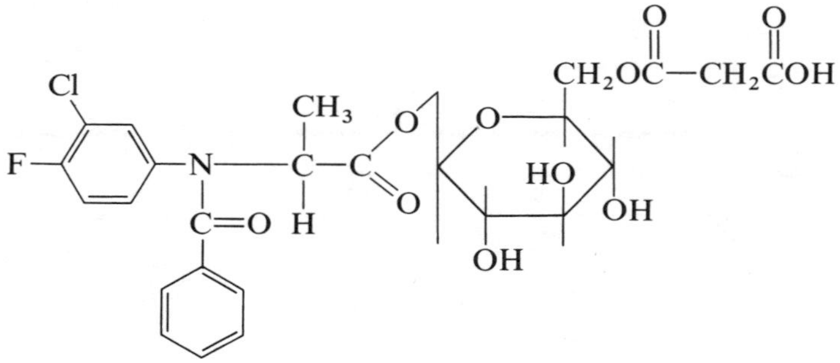

(14)

A detailed review of the 220 MHz n.m.r. spectra of cyanogenic glycosides should also be very useful for the recognition of aglycone stereochemistry and the identification of cyanogenic glucosides (Turczan *et al.*, 1978).

Infrared spectroscopy (i.r.)   Few published papers have described extensive use of i.r. spectroscopy for the characterization of intact glycoside conjugates, although Still (1968) has recorded the i.r. spectrum of a glucosyl conjugate of

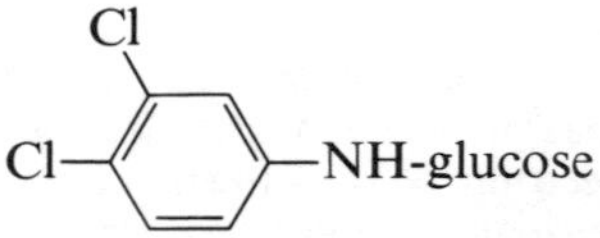

(15)

3,4-dichloroaniline (15) as its TMSi derivative. This was achieved by collecting the g.c. effluent directly on KBr discs, and comparing the spectra of the isolated metabolite with that of authentic standard material. However, as a technique it has not been widely used for intact conjugates, although for analysis of the aglycone moieties of conjugates after hydrolysis, where the spectra are less complex, it has received some interest (Jordan *et al.*, 1975; Frear and Swanson, 1975; Kamimura *et al.*, 1974). Current developments in Fourier Transform i.r. spectrometry (Krishnan *et al.*, 1979) which is capable of analysing ng amounts of chemicals may well be of some value in conjugate identification.

Chemical and enzymatic analysis

In many instances, the use of specific chemical derivatization techniques will make the conjugate more amenable to instrumental analysis. The presence of polar functional groups, e.g. hydroxyl, carboxyl or carbonyl, will often lead to poor chromatography. However, simple standard chemical reactions, e.g. methylation, acetylation or silylation, can be of value, and reference to many of the publications cited in this review will illustrate the use of such reactions. Many of these basic chemical reactions can be carried out in a variety of solvents, depending on the nature of the conjugate. In addition to effectively identifying polar functional groups, additional clean-up will often also be achieved. This results from a reduction in polarity which facilitates partition into a less aqueous solvent (More *et al.*, 1978). Concomitant changes in volatility and chromatographic mobility often result, and the value of trimethylsilylation in mass spectrometry of glucosides has been discussed (Martinelli, 1980).

Chemical or enzymatic hydrolyses are widely used in the study of conjugates, but have the disadvantage of being destructive. This will often be a disadvantage where only a small amount of conjugate can be isolated. Nonetheless, chemical and enzymatic hydrolyses are widely used and on the basis of such reactions many metabolites are classed as 'conjugates' (see Table 2). It has already been pointed out that such criteria, in themselves, are insufficient for complete identification (Frear, 1976). While, in the context of a full structural identification, chemical and enzymatic hydrolysis reactions can be extremely valuable (Hodgson and Hoffer, 1977b; Croteau and Martinkus, 1979), it should be emphasized that the specificity of enzyme preparations used can vary widely. Thus tentative claims in the literature for 'glucose conjugates' on the basis of β-glucosidase hydrolysis and cochromatography of the aglycone with standards should not be regarded as acceptable practice.

Absolute proof of structure, in addition to characterization of the intact conjugate, should also include separate characterization of the aglycones and sugar groups. While discussion of procedures for identification of aglycones is outside the scope of this review, some recent trends in the analysis of the sugar residue will be discussed in the following section.

Characterization of the sugar moieties of conjugates

Proof of conjugate structure depends not only on characterization of the intact conjugate, but also on identification of the constituent sugar and aglycone. Traditionally, methods of sugar analysis, on a microscale, have involved a chromatographic method, particularly paper chromatography. However, recent advances in the application of sensitive spray reagents for sugars on t.l.c. plates (Bounias, 1980) and derivatization techniques to simplify g.l.c. or h.p.l.c. analysis (Lawson and Russell, 1980), have broadened the range of methods available.

In order to determine the sugar residue of the conjugated metabolites of 3-hydroxy-5-methyl isoxazole, Kamimura *et al.* (1974) used descending paper chromatography of the hydrolysed conjugate in ethyl acetate–pyridine–water (10 : 4 : 3, v/v) with silver nitrate as a spray reagent. Further, they formed trimethylsilyl derivatives of the sugar and compared g.l.c. retention times with silylated standard compounds. A combination of these two techniques revealed glucose as the sugar moiety. The linkage of the glucose on the 1-position to the aglycone was deduced from permethylation, g.l.c. of the derivative along with authentic penta-*O*-methyl-*β*-glucopyranoside and a negative reaction with aniline hydrogen phthalate on the parent metabolite. The *β*-configuration of the sugar linkage was confirmed by optical rotatory dispersion and by hydrolysis of the conjugate with *β*-glucosidase.

In a detailed study of the conjugated metabolite of perfluidone (16) Lamoureux and Stafford (1977) identified glucose as the principal sugar by silylation using pyridine–hexamethyldisilazine–trimethylchlorosilane, followed by g.l.c. of the silylated products.

(16)

In two studies on the conjugation of pyrethroid metabolites in plants, More *et al.* (1978) and Wright *et al.* (1980) used descending paper chromatography with sodium acetate impregnated paper. After elution with ethyl acetate–isopropanol–water (65 : 23.5 : 11.5, v/v), the monosaccharides were visualized with 4-aminobenzoic acid and orthophosphoric acid.

In a detailed study of the structure of (+)-neomenthyl-*β*-D-glucoside in peppermint (*Mentha piperita*), Croteau and Martinkus (1979) reduced the aqueous residue of the conjugate hydrolysis products with sodium[³H]borohydride. The [³H]alditol formed by reduction of the corresponding aldose was shown to be a single product by t.l.c. on silica gel-G and cellulose. Oxidation of the [^{3}H]alditol with nitrous acid to the corresponding saccharic acid indicated the presence of a [1-^{3}H]alditol since most of the ^{3}H was recovered as [^{3}H]water. Further proof of structure was obtained by radio-g.l.c. of the trimethylsilylated and acetylated [^{3}H]alditol with authentic reference compounds. The presence of glucitol was finally proved by recrystallization of the [^{3}H]alditol with glucitol to constant specific activity followed by recrystallization of the hexabenzoate derivative of glucitol to constant specific activity. On the basis of this evidence, it was concluded that the labelled neutral carbohydrate was [1-^{3}H]glucitol which must have been derived from sodium-[^{3}H]borohydride reduction of glucose, obtained by hydrolysis of the glycoside.

Such detailed studies are often impossible in pesticide conjugate studies owing to lack of pure material. However, the previous pesticide studies give some indication of the approach necessary when working with very small amounts of conjugates.

Undoubtedly a variety of techniques now exist to study monosaccharide structure, and the feasibility of using these will depend on the amount of conjugate available. An excellent review (Laker, 1980), outlines a general approach for the identification of neutral sugars and sugar alcohols in biological fluids and shows the value of derivatization followed by g.l.c. analysis.

Developments in h.p.l.c. technology, in particular detection systems, are likely to be of value in conjugate studies and recently Wells and Lester (1979) have made use of reverse phase h.p.l.c. with a moving wire detector to separate a series of acetylated oligosaccharides. Binder (1980) has also made a study of the applicability of a chemically bonded amine stationary phase and a comparison of refractive index and u.v. detectors for monosaccharide determination. Games and Lewis (1980) have assessed the value of on-line l.c./m.s. for the characterization of mono- and disaccharides and glycosides.

Alternative approaches using immobilized enzymes have also been reported. Miller et al. (1977) described the use of immobilized peroxidase to determine the peroxide produced in the oxidation of glucose by an immobilized glucose oxidase, and were able to detect glucose at the mg kg^{-1} level.

Lubrano and Guilbault (1978) constructed amperometric enzyme electrodes by cross-linking bovine serum albumin and glucose oxidase, and carried out a detailed study of the effect of membrane thickness and enzyme activity.

Undoubtedly alternative methods for sugar analysis will become available in the future, although it is likely that for conjugate studies, the use of chromatography will continue to be of primary importance.

Conclusions

In this section, an attempt has been made to highlight the principal techniques which have been available for the characterization of pesticide conjugates. This characterization and identification process can be likened to a chemical jigsaw puzzle where structural information from a range of techniques has to be carefully assessed and pieced together. The amount of conjugate available will inevitably depend on the method of preparation and the care exercised in clean-up of the intact conjugate.

For the newcomer to pesticide conjugate analysis a brief list of some recent references which illustrate the use of a wide range of techniques both instru-mental and chemical, in the characterization of pesticide conjugates is provided. A study of these projects should provide some insight into the current applica-tion of analytical and chemical methods to conjugate identification: diphenamid (Hodgson and Hoffer, 1977b), 3-hydroxy-5-methylisoxazole (Kamimura *et al.*, 1974), flamprop (Dutton *et al.*, 1976), perfluidone (Lamoureaux and Stafford, 1977), oxamyl (Harvey *et al.*, 1978), 3-phenoxybenzyl alcohol (More *et al.*, 1978 and chloramben (Frear *et al.*, 1978).

ANALYSIS OF FIELD RESIDUES OF CONJUGATES

The ultimate aim of a pesticide metabolism study is to obtain an understanding of the likely residue of metabolites that will occur in actual field crop samples. For this a residue analysis procedure for the metabolite is required. Although it is now quite common to develop residue methods for metabolites, only a rather limited number of procedures have been published for pesticide conju-gates. While this may indicate an assumption that metabolites and their con-jugates are similar toxicologically, it also reflects the difficult technical challenge in developing such methods.

Examples of residue analysis procedures for conjugates include the following:

1. Maleic hydrazide (Newsome, 1980).
2. Metribuzin, as the desaminodiketo derivative (Thornton and Stanley, 1977).
3. Carbofuran, as phenolic derivatives (Cook *et al.*, 1977).
4. Dimethoate, as N-hydroxymethyl analogue (Stellar and Brand, 1974).
5. Chlorthiamid, as 3-hydroxy-2,6-dichlorobenzamide (Beynon and Wright, 1972b).
6. Dichlobenil, as hydroxy benzonitrile (Beynon and Wright, 1972b).
7. Monocrotophos, as N-hydroxymethyl analogue a. (Porter, 1967) b. (Beynon *et al.*, 1968).
8. Chlorfenvinphos, as ethan-1-ol derivative (Beynon *et al.*, (1968).
9. Tetrachlorvinphos, as ethan-1-ol derivative (Beynon *et al.*, (1970).
10. Benzoylprop-ethyl, as 'benzoylprop' acid (Wright and Mathews, 1976).
11. Flamprop-isopropyl, as 'flamprop' acid (Bosio *et al.*, 1981).

All of these procedures involved extracting the conjugates from the plant matrix, hydrolysis, usually to the aglycone, and clean-up of the extract prior to analysis, usually by g.l.c. Usually an acid hydrolysis was used (cases 3, 4, 5, 7, 8, 9) although β-glucosidase was recommended once (case 1) and boiling with water was preferred in case 2. In three procedures (cases 7b, 10, 11) the hydrolysis was combined with a derivatization: in 7b the monocrotophos conjugate was treated with s-butylthiol in hydrochloric acid to convert the conjugate directly into a thioether, and in 10 and 11 the 5% v/v sulphuric acid in methanol converts the conjugate directly into the methyl ester of the acid. It is worth considering analysis of the intact glycoside either by h.p.l.c. or after derivatization by acetylation or TMS, to increase volatility (Martinelli, 1980). Extensive analysis has been carried out for IAA both free and bound, and owing to its instability, ^{14}C and deuterated IAA were used to measure the recoveries from conjugate, an alkaline hydrolysis procedure being used (Bandurski, 1979).

Development of methods for the determination of conjugated residues in crops is normally considerably more laborious than for free residues. Conjugates are normally very polar and are extracted in solvents such as acetonitrile + water (70 + 30 v/v) (cases 10, 11), methanol (7), aqueous acid (4), (3) or hot water (2) in contrast to usual residue practice which is to use a solvent of minimal polarity compatible with good extraction of a non-polar pesticide. As a result, the more polar extract contains more coextractives than usual. Clean-up has normally been by a solvent wash to remove non-polar coextractives followed by column chromatography on Florisil (cases 2, 3, 9) or alkaline alumina (8, 9, 10) or Amberlite XAD-2 and Dowex resins (1). In a few cases no separate clean-up was needed as the final chromatography gave sufficient separation. This occurred for dimethoate-treated grapes (case 4) and for various monocrotophos-treated crops (7). However, this is a problem area where improvements are still likely.

There are also difficulties in establishing the extraction procedure. Since the conjugates are best biosynthesized in the crop to obtain a realistic sample their concentration is initially unknown. Accordingly the study is best carried out using ^{14}C-radiolabelled substrate allowing measurement of both extracted and unextracted residue. Many of the examples quoted used this procedure (cases 2, 5, 7, 8, 9, 10, 11). Work with radiolabelled materials has indicated that at increased contact times, the residue becomes less readily extractable, more polar and less open to hydrolysis and fractions of different polarity can differ widely in their reaction with enzymes such as commercial β-glucosidase. A novel approach to solving these problems would be the use of an immunological procedure such as radioimmunoassay as has been applied to S-bioallethrin (Wing and Hammock, 1979). Potentially this offers the necessary sensitivity and specificity but although it has been applied to free pesticides, no application to conjugate determination is yet known. The application of immunological techniques to pesticide analysis has been recently reviewed (Hammock and Mumma, 1980).

Residue concentrations of pesticide conjugates in crops

Some caution is necessary when considering residue concentrations, in extending the results of ^{14}C-radiolabelled studies to the field situation and this applies to conjugate residues as well as to the free compounds. The conjugates are derived from compounds which in the field are much more open to loss by volatilization and wash-off by rain. For example, although conjugates derived from monocrotophos could be detected using labelled studies, Beynon et al. (1973a) observed that while the residue procedure had been applied to some 16 types of field-treated samples, no residues of the neutral conjugates of the N-hydroxymethyl compound were detected in most crops, although small residues (0.05–0.35 mg kg^{-1}) were occasionally found in carrots, olives and oranges. Similarly, no residues at the 0.05 mg kg^{-1} sensitivity limit of the conjugates of dimethoate were found in grapes at 28 days from application (Stellar and Brand, 1974). Small residues of the conjugates of the ethan-1-ol derived from tetrachlorvinphos (but not chlorfenvinphos) have been reported (Beynon et al., 1968, 1970).

More extensive data are available for the conjugates of 'benzoylprop', the acid derived from benzoylprop-ethyl. Bosio et al. (1981) have shown that small but measurable residues of the conjugates are often detectable in the straw of field-treated wheat, the size of the residue increasing when the application was later than the recommended timing (Feekes scale G-I). Residues of benzoylprop-ethyl and 'benzoylprop' acid were not usually detected in grain following normal commercial applications in the United Kingdom, but were sometimes detected in crops from applications in other countries. When the applications were deliberately made at later than the recommended timings, at Feekes Scale I–J or later, conjugate residues could be obtained, and using a sample treated at stage M, it was shown that much of the conjugate residue initially present (0.51 mg kg^{-1}) was concentrated in the bran fraction (2.70 mg kg^{-1}) on milling. No conjugate residue was detected in the white flour or shorts ('fine offal'). The conjugates were also detected in brown bread baked from the flour.

The processing of crops into commercial products can also lead to decreased residues elsewhere. Thus Beynon and Wright (1972b) indicated that while chlorthiamid applications to vines can lead to detectable residues of the conjugates of the hydroxybenzamide, these are largely removed by the pressing or vinification process. Furthermore, Bosio et al. (1981) also suggest that although conjugates of 'flamprop' (from flamprop-isopropyl) may be detectable in some treated barley samples, they were not detectable in beer made from one of them.

THE NATURE AND IDENTITY OF SUGAR CONJUGATES OF PESTICIDES

The breakdown of numerous pesticides and other xenobiotics has been extensively studied in plants. However, although the subject has been well documented a listing of the conjugates formed cannot readily be found by a computer search of the relevant literature. This may well be a result of inconsistencies in nomenclature since conjugates are often recorded merely as 'polar products' or 'sugar derivatives'.

Primarily for this reason, a range of pesticides forming conjugates is recorded in Table 2. It may well be of value in locating original publications on pesticide conjugates, although the intention is not to quote the original paper but rather to cite a relevant study.

A study of Table 2 can lead to some general observations.

1. Alcohol groups are usually conjugated readily. They are usually derived from the parent pesticide by oxidation or hydrolysis, e.g. chlorfenvinphos to the ethan-1-ol (Beynon et al., 1968).

   An exception to this is triphenylcarbinol (formed from trifenmorph), where conjugates were largely detected from phenolic oxidation products (Beynon and Wright, 1967). In this instance, the barrier to alcohol conjugation may be steric hindrance or the high lipophilicity of triphenylcarbinol.
2. Phenolic groups may be formed by hydrolysis, e.g. carbaryl (Kuhr and Casida, 1967) or oxidation of an aromatic group. While phenols are readily and effectively conjugated, the rate of oxidation of an aromatic ring to its corresponding phenol may well be slow and will be subject to the effects of other ring substituents.
3. Carboxylic acid groups will also readily conjugate to form sugar esters. These are more susceptible to hydrolysis than the above classes and may well have been overlooked in some studies. As with phenols, caution should be exercised when predicting the existence of conjugates for highly substituted carboxylic acids, e.g. 2,3,6-trichlorobenzoic acid and trichloroacetic acid (Spitznagle et al., 1968).

It is interesting that while triazine–amino acid conjugates have been isolated, no case of triazine–glucose conjugation has apparently been reported, other than conjugation through a ring substituent (17) (Ciba-Geigy, 1975).

$$SCH_3$$

$$(CH_3)_2HCN(H) \quad NHCH_2CH_2CH_2O\text{-glucose}$$

(17)

BIOCHEMICAL ASPECTS OF
PESTICIDE GLYCOSIDE FORMATION

The previous sections have outlined the enormous range of pesticides which are conjugated with sugars when applied to plants. With modern analytical techniques the precise nature of the sugar, the aglycone and even the stereochemistry of the joining bond can often be elucidated. However, the enzymes responsible

Table 2 Selected examples of pesticides for which the formation of conjugates has been reported

Pesticide	Aglycone	Conjugate or sugar present	Notes on techniques	Reference
Amitrole	parent	glucoside	—	Frederick and Gentile (1960)
Abscisic acid	parent	β-D-glucoside	reaction with methanol under neutral or base conditions	Milborrow and Mallaby (1975)
Aldicarb	2-mesyl-2-methylpropan-1-ol and 2-mesyl-2-methylpropionitrile	—	β-glucosidase and 2N-hydrochloric acid hydrolysis	Rouchaud et al. (1980)
4-Aminopyridine	parent	sugar conjugates	nutrient culture preparation	Starr and Cunningham (1975)
Benzoylprop-ethyl	'benzoylprop' acid	sugars including glucose	β-glucosidase hydrolysis cochromatography and	Beynon et al. (1974)
2-secButylphenyl-N-methylcarbamate	several	β-glucose	β-glucosidase hydrolysis	Ogawa et al. (1976)
Carbaryl	N-hydroxymethyl, 4-hydroxy, 5-hydroxy and 5,6-dihydro-5,6-dihydroxy metabolites	glucose	β-glucosidase hydrolysis	Kuhr and Casida (1967)
Carbofuran	3-hydroxycarbofuran	glucosides and oligosaccharides 'conjugates'	g.l.c. of acetate; enzyme hydrolysis	Knaak et al. (1970)
Chlomethoxynil	hydroxy and amino metabolites	β-glucose	—	Niki et al. (1976)
Chloramben	parent	α-D-glucose, (both N- and O-ester glucosides occur)	—	Frear et al. (1978)
Chlorfenprop-methyl	polyhydroxylated aromatic metabolites	β-D-glucosides	β-glucosidase hydrolysis	Pont and Collet (1980)
Chlorfenvinphos	2,4-dichlorophenyl-ethan-1-ol	glucose	—	Beynon et al. (1968)
2-Chloro-4,5-xylyl methyl-carbamate and 2-Isoprop-oxyphenyl methyl carbamate	various including N-hydroxymethyl metabolites	glucose	β-glucosidase hydrolysis	Kuhr and Casida (1967)

Compound	Metabolite / product	Conjugate	Method	Reference
Chlorpropham	2- and 4-hydroxy-chlorpropham	glucose and poly-saccharides	h.p.l.c. and g.l.c., β-glucosidase, cellulase	Still and Mansager (1975)
Chlorthiamid	3-hydroxy-2,6-dichloro-benzamide	glucose	cochromatography and β-glucosidase hydrolysis	Beynon et al. (1972b)
Chlortoluron	4-hydroxymethyl and 4-carboxy analogues and demethyl metabolites	glucose	cochromatography and β-glucosidase hydrolysis	Gross et al. (1979)
Cisanilide	primary aryl and alkyl oxidation products	O-glucosides	β-glucosidase	Frear and Swanson (1975)
Cypermethrin	3-phenoxybenzoic acid	glucose and glucosylpentose	see techniques section	More et al. (1978)
Cypermethrin	trans-2[2,2-dichlorovinyl]-3,3-dimethylcyclopropane carboxylic acid	glucose and glucosylpentose	see techniques section	Wright et al. (1980)
2,4-D	2,4-D and hydroxylated metabolites	glucose	β-glucosidase hydrolysis	Feung et al. (1976)
(also related compounds, e.g. 2,4-DB	—	47% glucosides; also cell wall polysaccharides	enzymic studies	Smith (1979)
Decamethrin	several	glucose, and mainly more polar polysaccharides	β-glucosidase hydrolysis	Ruzo and Casida (1979)
Diamidofos	phenol	β-D-glucoside and other glycoside	cochromatography with intact conjugate	Meikle (1977)
Diazinon	—	water soluble compounds	yields of water soluble compounds depend on nutrient composition	Kunstman and Lichenstein (1979)
2-(2,4-dichlorophenoxy)iso-butyric acid	parent	β-D-glucoside	acetylation, m.s.	Gross and Schütte (1980)
Dicamba	several	glucose	—	Chang and van den Born (1971)
Dichlobenil	hydroxylated dichlobenil	plant polymers	—	Verloop (1972)
3',4'Dichloro-propionanilide	3,4-dichloroaniline	glucose, also xylose, fructose and lignin	i.r., isotope dilution with standard conjugate	Yih et al. (1968)
Diclofop-methyl	parent and ring-hydroxylated metabolites	glucosyl ester and ether	—	Dusky et al. (1980)

(continued)

Table 2 (*continued*)

Pesticide	Aglycone	Conjugate or sugar present	Notes on techniques	Reference
Dicrotophos	N-hydroxy metabolite	possibly glucose	—	Bull and Lindquist (1964)
Dimethoate	N-hydroxy metabolite	glucose	residue analysis	Steller and Brand (1974)
4-Dimethylamino-3,5-xylylmethylcarbamate	phenolic compounds	conjugates including lignin complexes	—	Williams et al. (1964)
Dimethyldithiocarbamate, Na⁺ salt	parent	β-glucoside	cochromatography with standard conjugate	Kaslander et al. (1961)
Dimethyl-4-(methylthiophenyl)phosphate	phenolic compounds	glucoside	β-glucosidase hydrolysis	Bull and Stokes (1970)
Dinoben	chloramben	N-glucoside	comparison with standard N-glucosyl chloramben	Colby (1966)
Dinobuton	hydrolysis reduction products	glucose	β-glucosidase hydrolysis	Bandal and Casida (1972)
Diphenamid	parent and N-methylol derivatives	disaccharide and acidic glucoside	cell culture	Davis et al. (1978)
Epronaz	N-ethyl-N-propyl-3-propylsulphonyl-1,2,4-triazole-1-carboxamide or 3-propylsulphonyl-1,2,4-triazole (unidentified)	water soluble compounds	—	Mellis and Kirkwood (1980)
O-Ethyl O-[4-methylthiophenyl] S-propyl phosphorodithioate	p-methylthiophenol	glucose	β-glucosidase hydrolysis and cochromatography with authentic aglycone	Bull et al. (1976)
Fenitrothion	3-methyl-4-nitrophenol	glucoside	—	Hosokawa and Miyamoto (1974)
Fenvalerate	various	glucose	β-glucosidase hydrolysis	Ohkawa et al. (1980)
Flamprop-isopropyl	flamprop	glucose, hemimalonyl glucoside	h.p.l.c., m.s.	Dutton et al. (1976)

Compound	Metabolite	Conjugate	Method	Reference
Fluchloralin	'metabolite V'	—	not sensitive to β-glucosidase or hesperidinase	Marquis et al. (1979)
Fluorodifen	p-nitrophenol, and 2-amino-4-trifluoromethylphenol	mixed conjugates	—	Shimabakuro et al. (1973)
Hexadecylcyclopropane carboxylate	mixed cyclopropane carboxylic acids	—	alkaline hydrolysis	Quistad et al. (1978)
3-Hydroxy-5-methyl isoxazole	parent and 5-methyl-4-isoxazolin-3-one	β-D-glucose	isolated as crystals for n.m.r., i.r., u.v. ORD and hydrolysis studies	Kamimura et al. (1974)
Isopropyl-3-chloro-carbanilate	isopropyl-5-chloro-2-hydroxycarbanilate	O-glucosidase	β-glucosidase hydrolysis, acetylation, g.c./m.s.	Still and Mansager (1972)
3-Isopropylphenyl methylcarbamate	N-hydroxymethyl and hydroxypropyl metabolite	glucose	—	Kuhr and Casida (1967)
Isoxathion	3-hydroxy-5-phenyl-isoxazole and 2 other aglycones	β-D-glucoside	cochromatography with standards of conjugates	Ando et al. (1975)
Kitazin	possible m-hydroxy analogue	'water-soluble products'	—	Tomizawa and Uesugi (1972)
Lindane	tetrachlorophenol and pentachlorophenol	—	—	Kohli et al. (1976)
Maleic hydrazide	parent	glucose	acetylation, TMS, m.s., β-glucosidase hydrolysis	Frear and Swanson (1978)
MCPA	MCPA and hydroxylated metabolites	glucose	—	Collins and Gaunt (1971)
Mephosfolan	methylol metabolites	glucosides	β-glucosidase hydrolysis	Zulalian and Blinn (1977)
Metflurazone	—	polar products	—	Motooka et al. (1977)
Methabenzthiazuron	N-methylol metabolite	glucose	—	Collet and Pont (1974)
Methidathion	possibly des-methyl metabolite	conjugate	—	Casida et al. (1969)
Metribuzin	metribuzin or one of its metabolites	N-glucose	g.l.c. of glucose, cochromatography with intact standard	Smith and Wilkinson (1974)

(continued)

Table 2 (*continued*)

Pesticide	Aglycone	Conjugate or sugar present	Notes on techniques	Reference
Monocrotophos	*N*-hydroxy metabolite	glucose and other sugars	—	Beynon and Wright (1972a)
Monolinuron	*N*-methylol metabolite	glucose	breakdown route reversible	Schuphan (1974)
Monuron	2-hydroxy metabolite	glycosides, including glucose triglucoside	β-glucosidase and hesperidinase hydrolysis	Frear and Swanson (1974)
Naphthalene acetic acid	parent		—	Schlepphorst and Barz (1979)
Oxamyl	oximino metabolites	glucose and possible polymers	—	Harvey *et al.* (1978)
Perfluidone	3-hydroxy metabolite	β-D-glucoside	β-glucosidase hydrolysis	Lamoureux and Stafford (1977)
Permethrin	various	conjugates	—	Gaughan and Casida (1978)
O-Phenyl-O′-(4-nitrophenyl)methylphos-phonothionate	*p*-nitrophenol	probably glucose	emulsin hydrolysis	Marco and Jaworski (1964)
Phosalone	benzoxazolone metabolite	glycoside	—	Colinese and Terry (1968)
Propanil	3,4-dichloraniline	glucosamine (also lignin)		Hodgson (1971)

Compound	Metabolite	Conjugate	Method	Reference
Propham	3-hydroxylated aromatic compounds	glucose		Zurqiyah et al. (1976)
Pyrazon	pyrazon	glucosamine	β-glucosidase, hesperidinase	Stephenson and Ries (1969)
Salithion	phenolic compounds	glucose	cochromatography with standard sample	Mihara and Miyamoto (1974)
Siduron	hydroxy metabolites	sugar conjugates	—	Jordan et al. (1975)
2,4,5-T		'polar products'	hydrolysis with β-glucosidase and hesperindase	Slife et al. (1962)
Temephos	phenolic compounds	glucosides	β-glucosidase hydrolysis	Blinn (1968)
Tetrachlorvinphos	2,3,5-trichlorophenyl-ethan-1-ol	glucose, possibly malonyl glucoside	—	Beynon et al. (1970)
Tirpate	polar metabolites of tirpate not characterized	conjugates	—	Hill and Krieger (1975)
Trichlorfon	unidentified metabolite	glucoside	β-glucosidase hydrolysis	Bull and Ridgway (1969)
2,4,6-Trichlorophenol-4'-nitrophenyl ether	2,4,6-trichlorophenol-4'-aminophenyl ether	N-glucoside	cochromatography with authentic conjugate	Shimotori and Kuwatsuka (1978)
Trifenmorph	triphenylcarbinol and hydroxy metabolites	conjugate	β-glucosidase hydrolysis	Beynon and Wright (1967)

for catalysing the formation of such sugar conjugates are very rarely examined in a pesticide metabolism study. This has resulted in a dearth of knowledge on the mechanisms of pesticide glycoside formation. Information available in the literature on this subject has been recently reviewed by Lamoureux and Frear (1979). A large proportion of the information presented in this section will refer to naturally occurring plant glycosides and glucose esters which have been studied from this point of view in more detail.

The source and nature of the sugar

It is clear from Table 2 that the most commonly detected sugar in pesticide conjugate formation is glucose. Glucoside biosynthesis requires the glucose to be supplied as a derivative which will release energy upon hydrolysis. Most commonly uridine diphosphate glucose (UDPG) is the sugar donor which itself is formed using energy from the hydrolysis of glucose-1-phosphate and uridine triphosphate:

$$UTP + glucose\text{-}1\text{-}P \rightleftharpoons UDPG + P\text{-}P$$

Transfer of glucose from existing conjugates (low energy glycosylation) is also observed occasionally (Satô and Hasegawa, 1972). Other nucleotide sugars, e.g. cytidine diphosphate glucose (CDPG), guanosine diphosphate glucose (GDPG) and thymidine diphosphate glucose (TDPG), are also found in plants and may be involved. The majority of studies of cell-free glucosyl transferase systems have used exogenous UDPG to stimulate glycoside formation. Where ADPG, CDPG or TDPG has been substituted the enzyme was often found to be very specific for UDPG (Hahlbrock and Conn, 1970; Larson, 1971; Chandra *et al.*, 1978). However, Barber (1962) using a cell-free system from mung beans found that TDPG could substitute for UDPG in the glucosylation of quercetin. A similar finding was reported by Grunwald (1978) while reviewing the glucosylation of sterols. TDPG was less effective than UDPG but all other nucleotide sugars assessed were ineffective. The K_m constants for UDPG in transglucosylation systems as determined by Lineweaver–Burk plots vary from 0.044 mM (Müller and Leistner, 1978) to 1.88 mM (Frear, 1967).

Glucose esters as well as glucosides are formed via UDPG. The biosynthesis of indole-3-acetic acid (IAA) glucose ester was demonstrated *in vitro* with an enzyme system from sweet corn (*Zea mays*) (Michalezuk and Bandurski, 1980).

Although glucose is the most common, a variety of other sugars are found in the endogenous glycosides of plants. For instance an IAA rhamnose ester is found in the anthers and carpels of *Peltophorum ferrugineum* flowers (Ganguly *et al.*, 1974). The rhamnose was characterized simply by hydrolysis of the conjugate followed by paper chromatography but no other sugars were found. In this study, no attempt was made to identify the rhamnose donor molecule.

Rhamnose also occurs in glycosidic linkage with flavonols (Barber, 1962) and anthocyanidins (Kamsteeg *et al.*, 1980, quercetin, a rhamnosyl glycoside) **(18)**). Other rather rare sugars are used in plant glycoside formation. The flavone glycoside graveobiosid B **(19)** contains the four-carbon sugar apiose

(18)   (19)

which is transferred to the glucoside via UDP-apiose (Ebel and Hahlbrock, 1977). A disaccharide conjugate of 3-phenoxybenzyl alcohol (a pyrethroid metabolite) was isolated from cotton leaves (More *et al.*, 1978). Here again the sugar attached to the aglycone was glucose but a pentose (mainly arabinose) formed the disaccharide conjugate. More information on the dynamics of formation of this particular disaccharide will be presented in the following section. More common, perhaps are the simple glucosides or the conjugates where the saccharide chain is built up from a number of successive glucosylations from UDPG.

The predominance of glucose in pesticide glycoside formation in plants referred to earlier may well reflect the availability of pre-existing UDPG within the cell. More likely perhaps is that the majority of glycosyl transferase enzymes are specific for UDPG although it appears that enzymes involved in the transfer of a second sugar to a monosaccharide conjugate will often use other sugars. It is important to bear in mind that all sugar conjugates are potential substrates for glycosidase enzymes and that a dynamic equilibrium exists between hydrolysis and glycosylation. This aspect will be discussed more fully in the next section. In addition subcellular compartmentilization of enzymes and substrates may play a vital role in the balance.

It is feasible that the administration of the xenobiotic could trigger the induction of the appropriate enzyme synthesis. The induction of UDPG-dependent glucosyl transferase enzymes has been studied in more detail in connection with the flavone–glycoside biosynthesis mentioned previously (Ebel and Hahlbrock, 1977). It was found that irradiation of cell suspension cultures of *Petroselinum hortense* produced a co-ordinated induction of two groups of enzymes involved in the flavone–glycoside biosynthesis. In the second group which appeared after a lag period of 4 h was the UDPG : flavonol 7-*O*-glucosyl transferase.

The acceptor specificity of glycosyl transferase

Early work by Yamaha and Cardini (1960) indicated than an *in vitro* glycosyl transferase system from wheatgerm would glucosylate several different poly-hydroxyphenols but would not utilize phenols or simple monohydroxylated phenols as substrates. More recently Storm and Hassid (1974) purified a soluble β-D-glucosyl transferase from germinating mung bean—*Phaseolus aureus*. They tested a variety of acceptors (see Table 3) and found that phenol and butan-1-ol were equally good substrates. They concluded that neither the acidities of the hydroxyl group nor the electronic properties of the molecules had much influence on their suitability but that their space-filling characteristics and length were the important parameters. The enzyme system exhibited a very high pH optimum (pH 10) for both phenol and butan-1-ol. This had been observed previously for other aliphatic acceptors (Hahlbrock and Conn, 1970), gluco-sylation of aromatic compounds usually has a lower pH optimum (Ibrahim and Boulay, 1980; Müller and Leistner, 1978). The relatively non-specific glucosy-lating activity described above could (Storm and Hassid, 1974) be due to the presence of several enzymes although evidence was presented to support the hypothesis that they had purified a single enzyme. This type of enzyme may be important in glucose conjugation of pesticides and their metabolites.

Recent work combining evidence from genetics as well as biochemistry (Heinsbroek *et al.*, 1980) has shown that at least some UDPG-dependent glu-

Table 3 Substrate specificity of glucosyl transferase,
(Storm and Hassid, 1974)

Acceptor	Relative reactivity* (%)
Phenol	100
p-Cresol	61
4-Chlorophenol	66
4-Bromophenol	62
4-Methoxyphenol	19
4-Ethylphenol	9
3-Ethylphenol	11
4-Propylphenol	4
4-Hydroxyacetophenone	14
Butan-1-ol	97
2-Methylpropan-1-ol	50
2-Methylpropan-2-ol	12
Methanol	9
Ethanol	12
Propan-1-ol	39
Pentan-1-ol	90
Hexan-1-ol	32

* Relative to phenol

cosyl transferase enzymes are highly specific for their substrates and also for the positions of glucosylation within the aglycone and the sugar. For instance, vitexin and isovitexin are glucosides which are similar in structure where the glucose is attached by a carbon–carbon bond to the aglycone. Vitexin and isovitexin are both found as 2″-O-glucosides i.e. the disaccharides (20) and (21).

(20)
R = glucose-$\beta 1 \rightarrow 2 - O$-glucose
$R_1 = H$

(21)
R = H
R_1 = glucose-$\beta 1 \rightarrow 2 - O$-glucose

However, there are two different glucosylating enzymes involved which are controlled by different genes. These enzymes have different pH optima and are specific for the different glucosides vitexin and isovitexin.

Ibrahim and Boulay (1980) examined the substrate specificity of a glycosyl transferase enzyme isolated from tobacco cell cultures. The enzyme which was partially purified (80-fold), was tested against a variety of coumarin and cinnamic acids. The highest activity was found with the coumarins—daphnetin (22) and esculetin (23), the products formed being identified as the corresponding 7-O-glucoside. Opening of the lactone ring or removal of one hydroxyl group resulted in a 50% drop in enzyme activity, methylation of the 6-OH position resulted in a 75% loss of activity, whereas methylation of the 7-OH position produced a complete loss in activity. This enzyme preparation although not absolutely pure still exhibited a remarkable substrate specificity. When [^{14}C]-glucose was used no radioactivity was in any position other than the 7-OH.

(22) Daphnetin R_1 = H, R_2 = OH, R_3 = OH
(23) Esculetin R_1 = OH, R_2 = OH, R_3 = H
(24) Scopoletin R_1 = OCH$_3$, R_2 = OH, R_3 = H
(25) Isocopoletin R_1 = OH, R_2 = OCH$_3$, R_3 = H

Pesticides with two or more functional groups can form a range of conjugates. For instance, chloramben (26) metabolism produces the glucose ester (12) or the N-glucoside (27) (Frear et al., 1978). Feeding these conjugates back to plants

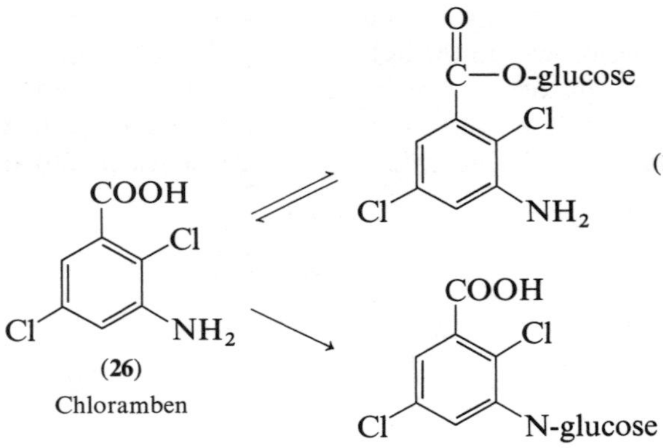

(12) Glucose ester

(26) Chloramben

(27) N-glucoside

demonstrated that the esterification is reversible whereas the N-glucosylation is probably irreversible. In addition, chloramben susceptible species have a reduced capacity to form the N-glucoside which means that more of the available chloramben forms the glucose ester which is apparently only a temporary detoxification. It would seem that from this work with susceptible and resistant species two different glucosyl transferases are at work.

The hydroxycinnamic acids are a group of bifunctional molecules which are found in plants. Fleuriet and Macheix (1980) have isolated both glucose esters and glycosides of several hydroxycinnamic acids from tomato fruit. The glucosides were found in higher concentrations than the esters throughout maturation and similar proportions were observed in a cell free system with exogenous substrate. Whether or not two enzymes are involved was not clarified by this work.

Stereospecificity of a glucosyl transferase is observed within the naturally occurring cyanogenic glycosides, a subject recently reviewed by Conn (1979). Sorghum contains the glucoside of the (S) enantiomer of p-hydroxy mandelonitrile (28), whereas the glucoside of the (R) isomer is found in several members

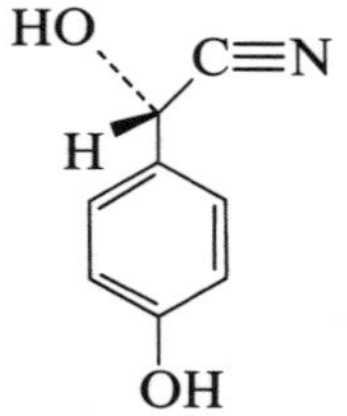

p-Hydroxy(S)-mandelonitrile (28)

of the bamboo family. Reay and Conn (1974) purified the glucosyl transferase from sorghum seedlings and analysed the glucoside formed when an enantiomeric mixture (RS) of the substrate was supplied to the enzyme system. Only dhurrin, the glucoside of the (S) isomer, was formed. However, the enzyme was less specific in other respects, i.e. (RS)-mandelonitrile itself was used equally well as substrate, hydroquinone and p-hydroxybenzyl alcohol also being gluco-sylated to a lesser extent (30–40%).

Another type of substrate specificity was exhibited by a D-glucosyl transferase enzyme isolated from germinating pea (Storm and Hassid, 1976). The enzyme transfers glucose from UDPG to an endogenous acceptor which appears to be a complex glycoside containing L-rhamnose, D-galactose and D-glucuronic acid, the aglycone being apparently phenolic in nature. The specificity of the enzyme was tested using phenol, hydroquinone, catechol, quercetin and several other phenols but enzyme activity was negligible with all of the compounds tested. These findings contrast sharply with the properties of the enzyme mentioned previously and isolated by the same group from germinating mung bean (Storm and Hassid, 1974).

The enzymes mentioned above have all been found in soluble cell fractions. This is likely to be the type of transferase involved in the glucosylation of most pesticides or their metabolites. However, the possibility exists that the more lipophilic xenobiotics are conjugated via membrane-bound enzymes. Sugar conjugates of a wide variety of lipids are found extensively in plants and recently some glucosylating enzymes have been studied. Quantin et al. (1980) have shown that a UDPG-dependent sterol-β-D-glucosyltransferase is present in maize coleoptiles and is mainly associated with the plasma membrane and endoplasmic reticulum. A similar enzyme was isolated in a particulate fraction from pea seedlings (Staver et al., (1978). This group found that the glucosyltransferase was specific for UDPG (ADPG, GDPG and CDPG were assessed) and the glu-cosylation reaction is inhibited by UDP. A uridine diphosphatase enzyme closely associated with the transferase, enhances glucosylation presumably by removing UDP.

The requirement for free sterol in sterolglucoside biosynthesis was established by Bush and Grunwald (1974) who incubated [4-^{14}C]cholesterol with a particu-late enzyme preparation from tobacco seeds and obtained radioactivity labelled cholesteryl glucoside. They also demonstrated that the enzyme was not sterol specific but Forsee et al. (1972) obtained contrasting results with a trans-glucosylase preparation from cotton seeds. Whether these enzymes would have a broad enough range to conjugate hydroxylated lipophilic compounds (i.e. 4′-OH cypermethrin (**29**)) is open to speculation.

(**29**)

Glucosyl esters of fatty acids have been found in rape pollen tissue. Chandra *et al.* (1978) have isolated a particulate cell fraction that catalyses the formation of glucosyl esters of fatty acids with carbon chain length C_{16} to C_{20}. The enzyme was specific for UDPG: other nucleotide diphosphate sugars were shown to be ineffective. Sterol glucosides are often acylated by esterification of fatty acids with the 6'-OH-position on glucose. Sugars are also incorporated into membrane bound proteins and lipids. There is evidence that plants, like animals, use poly-prenol phosphate sugar derivatives to transport sugars from the soluble nucleo-tide phosphates (e.g. UDPG) to the glycolipids and proteins of the cell membranes (Green and Northcote, 1979; Mellor *et al.*, 1980). The significance of these mechanisms in pesticide conjugation reactions is not clear.

Conclusions

Knowledge of enzymes involved in the formation of pesticide glycosides is limited and has been well reviewed by Lamoureux and Frear (1979). The aim of the above section has been to provide a summary of some of the recent bio-chemical literature, drawing examples mainly from natural products chemistry so that those interested in studying the mechanisms of pesticide glucoside forma-tion can have access to the current state of the art.

INTERCONVERSION REACTIONS OF PESTICIDE CONJUGATES

The majority of the conjugates reported in the studies listed in Table 2 were detected at one interval after application. It often appears that the conjugate fraction is a static 'sink' into which the product falls during detoxification. A limited number of studies suggest that this is not always the case, and that the conjugate fraction may in fact be undergoing continual reactions. Such an idea is in keeping with the presence of enzymes in the plant able to carry out such reactions, although an alternative possibility could be that conjugates are rendered unreactive by localization in certain parts of the cell.

Evidence that conjugates can persist in the field has partly been summarized elsewhere. An example is of the conjugates of 2-(2,4,5-trichlorophenyl)ethan-1-ol in apple fruit treated with tetrachlorvinphos. Here, Beynon and Edwards (1970) found little decrease over six weeks from application, the residues tending to increase as the last traces of the parent pesticide was metabolized. The analytical method used was developed intentionally to measure the total con-jugate present, and the results gave no indication of the nature of the com-ponents present.

It has often been commented that with increased time intervals from forma-tion, conjugate (as other) residues become less easy to extract, (e.g. Hodgson *et al.*, 1974), and become more polar (Frear *et al.*, 1977; Harvey *et al.*, 1978). This could be due to ripening of the plants in some cases. Thus it is worth considering some examples where interconversion has been demonstrated.

Glucosyl exchange

The bifunctional herbicide chloramben (26) is capable of conjugation at either the carboxylic or amino functions: the first identified was the N-glucosyl compound (Swanson et al., 1966). Subsequently, Frear et al., (1977) showed that the initial product was the glucose ester, which was labile in plants, giving both free chloramben and the relatively stable N-glucosyl compound. The biological significance of their findings has been discussed in the previous section.

Build-up of an oligosaccharide side chain

Build-up of an oligosaccharide side chain was postulated as a major reaction leading to more polar compounds in several of the now classical accounts and may well represent an important process. However, actual reported examples are still fairly scarce. The formation of a disaccharide such as a gentiobioside does not in itself demonstrate the transfer of a glucosyl group since the disaccharide group could also have been attached ready-made to the aglycone. However, it is likely that the gentiobiosides of o-chlorophenol reported by Miller (1940) and of hydroquinone, reported by Conchie et al., (1961) represent cases where a fresh glucosyl group has been transferred to the glucosyl group of a monosaccharide. This was more definite in the case of diphenamid (30) reported by Hodgson et al. (1974) where the products were analysed at intervals of 0.7 to eight days. The concentration of (32) rose and then fell owing to conversion into (33). It was also noticeable that the concentration of (31) remained finite, suggesting that it was in equilibrium with (32).

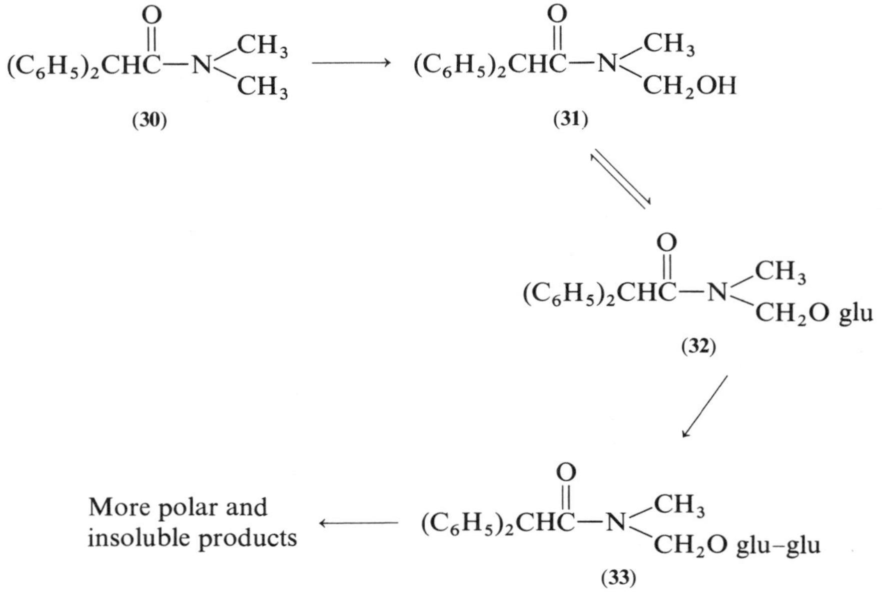

$$(C_6H_5)_2CHC(=O)-N\begin{array}{c}CH_3\\CH_3\end{array} \longrightarrow (C_6H_5)_2CHC(=O)-N\begin{array}{c}CH_3\\CH_2OH\end{array}$$

(30)   (31)

$$(C_6H_5)_2CHC(=O)-N\begin{array}{c}CH_3\\CH_2O\ glu\end{array}$$

(32)

$$\text{More polar and} \longleftarrow (C_6H_5)_2CHC(=O)-N\begin{array}{c}CH_3\\CH_2O\ glu-glu\end{array}$$
insoluble products

(33)

Another type of conversion became known as a result of studies with ^{14}C-labelled metabolites (11), (35) and (36) of the pyrethroid insecticide, cypermethrin (34). These were supplied as aqueous solutions to the cut stems of cotton and other leaves, and the sample was extracted after periods of up to 72 h. More *et al.* (1978) showed that (36) was initially converted into the glucosylester in high yield. Subsequently this was isolated and supplied to fresh leaves, whereon both free (36) and a new, more polar compound were obtained. The latter was separated into four components by h.p.l.c. and all of these gave both glucose and a pentose, apparently arabinose in three cases and xylose in the fourth. Such glucosylpentose derivatives of natural products are fairly well known but were novel as pesticide derivatives.

(34) Cypermethrin

(11)   (35)   (36)

Subsequently the acid (11) was shown by Wright *et al.* (1980) to be readily converted into mono- and disaccharides, again with glucose, arabinose and xylose. In addition, the monosaccharide fraction was partly acidic. Finally, Roberts and Wright (1981) showed that the alcohol (35) was converted into a similar mixture of mono- and disaccharides, with glucose linked to arabinose or xylose.

When these conjugates of (35) were fed back separately to the leaves in each case a mixture of aglycone, mono- and disaccharide was detected. The mono-saccharides of (11) and (35) underwent more reaction under these conditions than did the corresponding disaccharides. In the case of (35), these interconversions were also studied by analysing leaves at different intervals, the apparent rate constants being shown in Figure 3. The reversibility of the conjugation process may explain the detection of free aglycone in plants. Naphthylacetic acid has been shown by Schlepphorst and Barz (1979) to form a triglucoside (in plant cell suspension culture). Thus it may be that in cotton (Wright *et al.*, 1980)

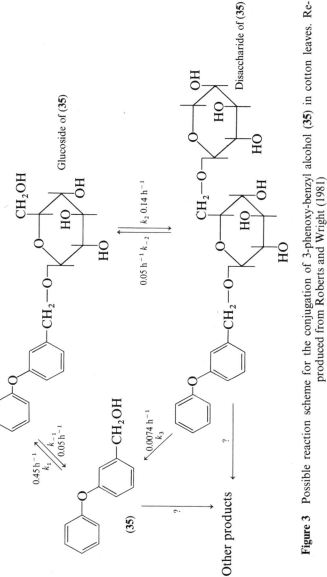

Figure 3 Possible reaction scheme for the conjugation of 3-phenoxy-benzyl alcohol (35) in cotton leaves. Reproduced from Roberts and Wright (1981)

the addition of the pentose group prevents further chain extension since the disaccharide produced lacks a primary alcohol.

Addition of non-sugar groups

The primary alcohol group on a glucosyl conjugate can, in principle, be blocked by other substituents, and at least one is known. This is the hemimalonyl group, which has been detected in conjugates derived from fluorodifen (Shimabakuro *et al.*, 1973), diphenamid (Davis *et al.*, 1978) and flamprop-methyl (Dutton *et al.*, 1976). These may represent only a fraction of those actually formed since the hemimalonyl group is easily lost during analysis and has little effect on the polarity of the monosaccharide. Thus the acidic fraction in **11** could be of this type, as could be the acidic conjugate detected from (2,4,5-trichlorophenyl)-ethan-1-ol (a tetrachlorvinphos metabolite) (Beynon *et al.*, 1973b). In general, an electrophoresis study on any conjugate is desirable. It has been suggested that the malonyl group was transferred after formation of the conjugate (Mumma and Hamilton, 1979). It has been shown that over longer periods in intact plants the monosaccharide conjugates of flamprop (and MCPA) become converted into more polar compounds, which are probably water soluble polysaccharides and hemicelluloses, without evidence of proteins or lignin (Pillmoor, 1981). Some hydrolysis to flamprop always occurred during the extraction.

Conversion with loss of sugar groups

Several of the papers referred to above illustrate the replacement of the glucosyl group with hydrogen upon hydrolysis. Examples of the replacement with methyl groups from methanolic solvents under the influence of transes-terases during extraction are also noted. These reactions influence the amounts of conjugate detected and possibly the first determines its biological activity. This was more certainly demonstrated by Zenk (1963) who reported several studies with IAA, α-naphthylacetic and benzoic acids, the acids being initially conjugated as glucosyl esters, but being subsequently converted in the plant into aspartates. This was particularly clearly shown for [^{14}C]IAA using leaf discs of *Hypericum hircinum* exposed for 4 h, followed by transfer to a solution of non-radioactive IAA for a further 20 h. The total radioactivity present in the discs varied little, but the Glu : Asp ratio fell from 1.6 : 1 to 0.75 : 1 between 6 h and 24 h from the beginning of the study.

The initial esterification represented a rapid reaction to form an energy rich bond in the ester. The conversion into the aspartate involves the formation of a less reactive compound. Other workers have extended these studies and these were reviewed by Bandurski (1979). In maize, little amino acid conjugation occurred, the main products being IAA-*myo*-inositols, IAA-*myo*-inositol-glucosides, IAA glucosides, and a high molecular weight 1,4-glucan conjugate;

similar glucoprotein conjugates have been detected in oats by Percival and Bandurski (1976). Such studies are very relevant to an understanding of pesticide behaviour. They show the types of conjugate which can form and emphasise that the conjugate is in equilibrium with the free acid, the proportions varying with growth conditions.

Replacement with malate (as well as aspartate and oligoglucosides) has been reported for various aromatic acids (Schlepphorst and Barz, 1979), and these authors comment that the plant and acid can both affect the nature of the conjugate formed. In the case of chlortoluron, a major metabolic route in young wheat was to the benzyl alcohol and its conjugates, without oxidation to the benzoic acid. However, as the wheat ages, the benzyl alcohol was converted into the benzoic acid by hydrolysis of the conjugates and oxidation of the free alcohols (Gross *et al.*, 1979). Where the aglycone is unstable, such a reaction can lead to decomposition, and this may be an important route for the removal of *N*-methyl via *N*-methylol groups which are unstable and decompose to free NH groups (Schuphan, 1974).

CONCLUSIONS

The foregoing account has emphasized the recent advances which have been made in the study of pesticide sugar conjugates in plants. This progress reflects not only the availability of more sophisticated analytical instrumentation but also an increasing awareness of the significance of this portion of the pesticide residue.

In the past, the emphasis has been on a study of free metabolites and, to a lesser extent, on the aglycones released by chemical or enzymatic hydrolysis of the conjugates. Studies of pesticide sugar conjugates have been somewhat neglected, often owing to lack of genuine interest, coupled with experimental difficulties associated with isolation and identification of intact conjugates. Such water-soluble metabolites were often dismissed on the assumption that conjugation was synonymous with detoxification and minimum biological significance. However, it is now apparent that sugar conjugates are not necessarily static end-products of metabolism, nor are they biologically inactive.

On the contrary, they are liable to cleavage by plant enzymes and are thus a potential source of aglycone. In addition, sugar conjugation normally leads to increased solubility in cell fluids and could therefore provide a means of transport to alternative sites in the cell or the plant for subsequent release of the aglycone. The limited data available from field studies suggest that residues of pesticide conjugates are present, but often at fairly low concentrations. However, recent developments in h.p.l.c. and g.c./m.s., in particular, have simplified the examination of such low concentrations of intact conjugates, and coupled with a growing awareness of the biochemical significance of conjugates, should ensure that interest will be maintained in this developing area of pesticide science.

REFERENCES

Ando, M., Iwasaki, Y., and Nakagawa, M. (1975) 'Metabolism of isoxathion, O,O-diethyl-O-(5-phenyl-3-isoxazolyl)phosphorothionate in plants', *Agr. Biol. Chem.*, **39**, 2137–2143.

Bandurski, R. S. (1979) 'Chemistry and physiology of conjugates of indole-3-acetic acid', in *Plant Growth Substances* (N. B. Mandava, Ed.), American Chemical Society Symposium Series No. 111, pp. 1–17.

Bandal, S. K., and Casida, J. E. (1972) 'Metabolism and photoalteration of 2-*sec*-butyl-4,6-dinitrophenyl (DNBP) herbicide and its isopropyl-carbonate derivative (Dinobuton acaricide)', *J. Agr. Food Chem.*, **20**, 1235–1245.

Barber, G. A. (1962) 'Enzymic glycosylation of quercetin to rutin', *Biochemistry*, **1**, 463–468.

Beckey, H. D., and Schulten, H.-R. (1975) 'Field desorption mass spectrometry', *Angew. Chem. Intl. Ed.*, **14**, 403–415.

Beynon, K. I., and Wright, A. N. (1967) 'Breakdown of the molluscicide *N*-tritylmorpholine in soil and in rice', *Bull. World Health Organization*, **37**, 65–72.

Beynon, K. I., and Edwards, M. J. (1970) 'Residues of tetrachlorvinphos and its breakdown products on treated crops. II. Residues on apples', *Pestic. Sci.*, **1**, 254–265.

Beynon, K. I., and Wright, A. N. (1972a) 'The breakdown of [^{14}C]monocrotophos insecticide on maize, cabbage and apple', *Pestic. Sci.*, **3**, 277–292.

Beynon, K. I., and Wright, A. N. (1972b) 'The fates of the herbicides chlorthiamid and dichlobenil in relation to residues in crops, soils and animals', *Residue Reviews*, **43**, 23–53.

Beynon, K. I., Edwards, M. J., Elgar, K. E., and Wright, A. N. (1968) 'Analysis of crops and soils for residues of chlorfenvinphos insecticide and its breakdown products', *J. Sci. Food Agric.*, **19**, 302–307.

Beynon, K. I., Edwards, M. J., and Wright, A. N. (1970) 'Residues of tetrachlorvinphos and its breakdown products on treated crops. I. Method development', *Pestic. Sci.*, **1**, 250–254.

Beynon, K. I., Elgar, K. E., Mathews, B. L., and Wright, A. N. (1973a) 'The analysis of crops to determine neutral conjugates of an *N*-hydroxymethyl derivative of monocrotophos insecticide', *Analyst*, **98**, 194–201.

Beynon, K. I., Hutson, D. H., and Wright, A. N. (1973b) 'The metabolism and degradation of vinylphosphate insecticides', *Residue Reviews*, **47**, 55–142.

Beynon, K. I., Roberts, T. R., and Wright, A. N. (1974) 'The degradation of the herbicide benzoylprop ethyl following its application to wheat', *Pestic. Sci.*, **5**, 429–442.

Binder, H. (1980) 'Separation of monosaccharides by high-performance liquid chromatography: comparison of ultraviolet and refractive index detection', *J. Chromatog.*, **189**, 414–420.

Blaschke, G. (1980) 'Chromatographic resolution of racemates', *Angew. Chem. Int. Ed. Engl.*, **19**, 13–24.

Blinn, R. C. (1968) 'Abate insecticide. Fate of O,O,O',O'-tetramethyl-O,O'-dithiodi-p-phenylene phosphorothioate on bean leaves', *J. Agr. Food Chem.*, **16**, 441–445.

Bosio, P. G., Cole, E. R., Mathews, B. L., Woodbridge, A. P., and Wright, A. N. (1981) 'The behaviour of residues of flamprop-isopropyl and its breakdown products in barley', *Pestic. Sci.*, (submitted for publication).

Bounias, M. (1980) '*N*-(1-naphthyl)ethylenediamine dihydrochloride as a new reagent for nanomole quantification of sugars on thin-layer plates by a mathematical calibration process', *Anal. Biochem.*, **106**, 291–295.

Brain, K. R., and Lines, D. S. (1977) in *Plant Tissue Culture and its Biotechnological Applications* (W. Barz, E. Reinhard, and M. H. Zenk, Eds.), Springer-Verlag, Berlin, pp. 197–203.

Bull, D. L., and Lindquist, D. A. (1964) 'Metabolism of 3-hydroxy-N,N-dimethylcrotonamide dimethylphosphate in cotton plants, insects and rats', *J. Agr. Food Chem.*, **12**, 310–317.

Bull, D. L., and Ridgway, R. L. (1969) 'The metabolism of trichlorfon in animals and plants', *J. Agr. Food Chem.*, **17**, 837–841.

Bull, D. L., and Stokes, R. A. (1970) 'Metabolism of dimethyl *p*-(methylthiophenyl)phosphate in animals and plants', *J. Agr. Food. Chem.*, **18**, 1134–1138.

Bull, D. L., Whitten, G. J., and Ivie, G. W. (1976) 'Fate of *O*-ethyl-*O*-[4-methylthiophenyl]-*S*-propyl phosphorodithioate [Bay NTN9306] in cotton plants and soil', *J. Agr. Food Chem.*, **24**, 601–605.

Bu'lock, J. D. (1964) in *The Biosynthesis of Natural Products*. (J. D. Bu'lock, Ed.), McGraw-Hill, London, p. 2.

Bush, B. P., and Grunwald, C. (1974) 'Sterol glycoside formation in seedings of *Nicotiana tabacum* L.', *Plant Physiol.*, **53**, 131–135.

Cairns, T., Froberg, J. E., Gonzales, S., Langham, W. S., Stamp, J. J., Howie, J. K., and Sawyer, D. T. (1978) 'Analytical chemistry of Amygdalin', *Anal. Chem.*, **50**, 317–322.

Casida, J. E., Ryskiewich, D. P., and Murphy, R. T. (1969) 'Metabolism of Supracide in alfalfa', *J. Agr. Food Chem.*, **17**, 558–564.

Chandra, G. R., Mandava, N., and Warthen, J. D., Jr. (1978). 'Uridine diphosphate glucose: fatty acid glucosyl transferase activity of rape (*Brassica napus* L.) anther tissue', *Biochim. Biophys. Acta*, **526**, 387–397.

Chang, F. Y., and van den Born, W. H. (1971) 'Translocation and metabolism of dicamba in tartary buckwheat', *Weed Sci.*, **19**, 107–112.

Ciba-Geigy (1975), cited by Esser, H. O., Dupuis, G., Ebert, E., Vogel, C., Marco, G. J., 's-Triazines' in *Herbicides, Chemistry, Degradation and Mode of Action* (P. C. Kearney and D. D. Kaufman, Eds.), Marcel Dekker, New York, pp. 129–208.

Colby, S. R. (1966) 'The mechanism of the selectivity of amiben', *Weeds*, **14**, 197–201.

Colinese, D. L., and Terry, H.-J. (1968) 'Phosalone—a wide spectrum organo-phosphorus insecticide', *Chem. Ind.*, **Nov. 2, 1968** 1507–1511.

Collet, G. F., and Pont, V. (1974) 'Distribution and metabolism of methabenzthiazuron in several plant species', *Weed Res.*, **14**, 151–165.

Collins, D. J., and Gaunt, J. K. (1971) 'The metabolism of 4-chloro-2-methylphenyl-oxyacetic acid in plants', *Biochem. J.*, **124**, 9P.

Conchie, J., Moreno, A., and Cardini, C. E. (1961) 'A trisaccharide glycoside from wheat-germ', *Arch. Biochem. Biophys.*, **94**, 342–343.

Conn, E. E. (1979) 'Biosynthesis of cyanogenic glycosides', *Naturwissenschaften*, **66**, 28–34.

Cook, R. F., Jackson, J. E., Shuttleworth, J. M., Fullmer, O. H., and Fujie, G. H. (1977) 'Determination of the phenolic metabolites of carbofuran in plant and animal matrices by gas chromatography of their 2,4-dinitrophenyl ether derivatives', *J. Agr. Food Chem.*, **25**, 1013–1017.

Croteau, R., and Martinkus, C. (1979) 'Metabolism of monoterpenes. Determination of (+)-neomenthyl-β-D-glucoside as a major metabolite of (−)-menthone in peppermint (*Mentha piperita*)', *Plant Physiol.*, **64**, 169–175.

Davidonis, G. H., Hamilton, R. H., and Mumma, R. O. (1980) 'Metabolism of 2,4-dichlorophenoxyacetic acid (2,4-D) in soybean root callus', *Plant Physiol.*, **66**, 537–540.

Davis, D. G., Hoerauf, R. A., Dusbabek, K. E., and Dougall, D. K. (1977) 'Isopropyl *m*-chlorocarbanilate and its hydroxylated metabolites: their effects on cell suspensions and cell division in soybean and carrot', *Physiol. Plant.*, **40**, 15–20.

Davis, D. G., Hodgson, R. H., Dusbabek, K. E., and Hoffer, B. L. (1978) 'The metabolism of the herbicide Diphenamid (N,N-dimethyl-2,2-diphenylacetamide) in cell suspensions of soybean (*Glycine max*)', *Physiol. Plant.*, **44**, 87–91.

Domir, S. C. (1978) 'Translocation and metabolism of injected maleic hydrazide in silver maple and American sycamore seedlings', *Physiol. Plant.*, **42**, 387–390.

Donald, W. W., and Shimabukuro, R. H. (1980) 'Selectivity of diclofopmethyl between wheat and wild oat: growth and herbicide metabolism', *Physiol. Plant.*, **49**, 459–464.

Dorough, H. W. (1979) 'Conjugation reactions of pesticides and their metabolites with sugars' in *Advances in Pesticide Science* (A. Geissbuhler, Ed.), Pergamon, Oxford, Part 3, pp. 526–536.

Dreifuss, P. A., Wood, G. E., Roach, J. A. G., Brumley, W. C., Andrejewski, D., and Sphon, J. A. (1980) 'Field desorption mass spectrometry of cyanogenic glycosides', *Biomed. Mass Spectrom.*, **7**, 201–204.

Dusky, J. A., Davis, D. G., and Shimabukuro, R. M. (1980) 'Metabolism of dichlofopmethyl in cell suspensions of diploid wheat (*Triticum monococcum*)', *Physiol. Plant.*, **49**, 151–156.

Dutton, A. J., Roberts, T. R., and Wright, A. N. (1976) 'Characterization of the acid conjugates of flamprop in wheat', *Chemosphere*, **3**, 195–200.

Ebel, J., and Hahlbrock, K. (1977) 'Enzymes of flavone and flavonol–glycoside biosynthesis', *Eur. J. Biochem.*, **75**, 201–209.

Fenselau, C., and Johnson, L. P. (1980) 'Analysis of intact glucuronides by mass spectrometry and gas chromatography-mass spectrometry', *Drug Metab. Disp.*, **8**, 274–283.

Feung, C. S., Hamilton, R. H., and Witham, F. H. (1971) 'Metabolism of 2,4-dichlorophenoxyacetic acid by soybean cotyledon callus tissue cultures', *J. Agr. Food Chem.*, **19**, 475–479.

Feung, C. S., Hamilton, R. H., and Mumma, R. O. (1975) 'Metabolism of 2,4-D: VII. Comparison of metabolites from five species of plant cell callus tissue cultures', *J. Agr. Food Chem.*, **23**, 373–376.

Feung, C. S., Hamilton, R. H., and Mumma, R. O. (1976) 'Metabolism of 2,4-dichlorophenoxyacetic acid: 10. Identification of metabolites in rice root callus cultures', *J. Agr. Food Chem.*, **24**, 1013–1015.

Feung, C. S., Loerch, S. L., Hamilton, R. H., and Mumma, R. O. (1978) 'Comparative metabolic fate of 2,4-dichlorophenoxyacetic acid in plants and plant tissue culture', *J. Agr. Food Chem.*, **26**, 1064–1067.

Fleuriet, A., and Macheix, J. (1980) 'Derivés glucosés des acides hydroxycinnamiques de la tomate: évolution au cours de la vie du fruit et formation *in vitro*', *Phytochem.*, **19**, 1955–1958.

Forsee, W. T., Lame, R. A., Chambers, J. P., and Elbein, A. D. (1972) in *Methods in Enzymology* (V. Ginsburg, Ed.), Academic Press, New York, Vol. 28, part B, pp. 478–482.

Frear, D. S. (1967) 'Purification and properties of UDP-glucose: arylamine *N*-glucosyl transferase from soybean', *Phytochem.*, **7**, 381–390.

Frear, D. S. (1976) 'Pesticide conjugates —glycosides', in *Bound and Conjugated Pesticide Residues* (D. D. Kaufman, G. G. Still, G. Paulson, and S. K. Bandal, Eds.), American Chemical Society Symposium Series No. 29, pp. 35–54.

Frear, D. S., and Swanson, H. R. (1974) 'Monuron metabolism in excised *Gossypium hirsutum* leaves: aryl hydroxylation and conjugation of 4-chlorophenylurea', *Phytochem.*, **13**, 357–360.

Frear, D. S., and Swanson, H. R. (1975) 'Metabolism of cisanilide (*cis*-2,5-dimethyl-1-pyrrolidinecarboxanilide) by excised leaves and cell suspension cultures of carrot and cotton', *Pestic. Biochem. Physiol.*, **5**, 73–80.

Frear, D. S., and Swanson, H. R. (1978) 'Behaviour and fate of [^{14}C]maleic hydrazide in tobacco plants', *J. Agr. Food Chem.*, **26**, 660–665.

Frear, D. S., Swanson, H. R., Mansager, E. R., and Wien, R. G. (1977) *Chloramben metabolism in plants: isolation and identification of a glucose ester*, 173rd ACS Meeting, New Orleans, La., 20.3.1977.

Frear, D. S., Swanson, H. R., Mansager, E. R., and Wien, R. G. (1978) 'Chloramben metabolism in plants: isolation and identification of glucose ester', *J. Agr. Food Chem.*, **26**, 1347–1351.

Frederick, J. F., and Gentile, A. C. (1960) 'Formation of the glucose derivative of amitrole under physiological conditions', *Physiol. Plant.* **13**, 761–765.

Games, D. E., and Lewis, E. (1980) 'Combined liquid chromatography mass spectrometry of glycosides, glucuronides, sugars and nucleosides', *Biomed. Mass Spectrom.*, **7**, 433–436.

Games, D. E., Hinter, P., Kuhnz, W., Lewis, E., Weerasinghe, N. C. A., and Westwood, S. A. (1981) 'Studies of combined liquid chromatography–mass spectrometry with a moving-belt interface', *J. Chromatog.* **203**, 131–138.

Ganguly, T., Ganguly, S. N., Sircar, D. K., and Sircar, S. M. (1974) 'Rhamnose bound indole-3-acetic acid in the floral parts of *Peltoplorum ferrugineum*', *Physiol. Plant.*, **31**, 330–332.

Gaughan, L. C., and Casida, J. E. (1978) 'Degradation of *trans*- and *cis*-permethrin on cotton and bean plants', *J. Agr. Food Chem.*, **26**, 525–528.

Gorbach, S. G., Kuenzler, K., and Asshauer, J. (1977) 'On the metabolism of HOE 23408 OH in wheat', *J. Agr. Food Chem.*, **25**, 507–511.

Green, J. R., and Northcote, D. H. (1979). 'Polyprenyl phosphate sugars synthesised during slime-polysaccharide production by membranes of the root cap cells of maize (*Zea mays*)', *Biochem. J.*, **178**, 661–671.

Gross, D., and Schütte, H. R. (1980) 'Metabolism of 2-(2,4-dichlorophenoxy)-isobutyric acid in seedlings of barley, wheat, rye and corn', *Biochem. Physiol. Planzen.*, **175**, 154–162.

Gross, D., Gutenkunst, H., Blaser, A., and Hambock, H. (1980) 'Peak identification in capillary gas chromatography by simultaneous flame ionisation detection and ^{14}C-detection', *J. Chromatog.*, **198**, 389–396.

Gross, D., Laanio, T., Dupuis, G., and Esser, H. O. (1979) 'The metabolic behaviour of Chlortoluron in wheat and soil', *Pestic. Biochem. Physiol.*, **10**, 49–59.

Grunwald, C. (1978) 'Sterol glycoside biosynthesis', *Lipids*, **13**, 697–703.

Hahlbrock, K., and Conn, E. E. (1970) 'The biosynthesis of cyanogenic glycosides in higher plants. I. Purification and properties of a uridine diphosphate glucose–ketone–cyanohydrin β-glucosyltransferase from *Linum usitatissimum*', *J. Biol. Chem.*, **5**, 917–922.

Hammock, B. D., and Mumma, R. O. (1980) 'Potential of immunochemical technology for pesticide analysis', in *Pesticide Analytical Methodology* (J. Harvey, Jr. and G. Zweig, Eds.), American Chemical Society Symposium Series No. 136, pp. 321–352.

Haque, A., Schuphan, I., and Ebing, W. (1978) 'On the metabolism of phenylurea-herbicides. X. Movement and behaviour of a glucoside conjugate in plant and soil', *Chemosphere*, **8**, 675–680.

Harvey, J., Jr. (1980) 'Modern high-performance liquid chromatography in pesticide metabolism studies', in *Pesticide Analytical Methodology* (J. Harvey, Jr. and G. Zweig, Eds.), ACS Symposia Series No. 136, pp. 1–14.

Harvey, J. Jr., Han, J. C-Y., and Reiser, R. W. (1978) 'Metabolism of oxamyl in plants', *J. Agr. Food Chem.*, **26**, 529–536.

Heinsbroek, R., van Brederode, J., van Nigtevecht, G., Maas, J., Kamsteeg, J., Besson, E., and Chopin, J. (1980) 'The 2″-O-glucosylation of vitexin and isovitexin in petals of *Silene Alba* is catalysed by two different enzymes', *Phytochem.*, **19**, 1935–1937.

Heirwegh, K. P. M., and Compernolle, F. (1979) 'Micro-analytic detection and structure elucidation of ester-glycosides', *Biochem. Pharmacol.*, **28**, 2109–2114.

Hill, J. E., and Krieger, R. I. (1975) 'Uptake, translocation, and metabolism of Tirpate in tobacco *Nicotiana tabacum*', *J. Agr. Food Chem.*, **23**, 1125–1129.

Hodgson, R. H. (1971) 'Influence of environment on the metabolism of propanil in rice', *Weed Sci.*, **19**, 501–507.

Hodgson, R. H., and Hoffer, B. L. (1977a) 'Diphenamid metabolism in pepper and an ozone effect. I. Absorption, translocation, and the extent of metabolism', *Weed Sci.*, **25**, 324–330.

Hodgson, R. H., and Hoffer, B. L. (1977b) 'Diphenamid metabolism in pepper and an ozone fumigation effect. II. Herbicide metabolite characterisation', *Weed Sci.*, **25**, 331–337.

Hodgson, R. H., Dusbabek, K. E., and Hoffer, B. L. (1974) 'Diphenamid metabolism in tomato: time course of an ozone fumigation effect', *Weed Sci.*, **22**, 205–210.

Hosokawa, S., and Miyamoto, J. (1974) 'Metabolism of ^{14}C-Sumithion, O,O-dimethyl-O-(3-methyl-4-nitrophenyl)phosphorothioate in apples', *Botyu-Kogaku*, **39**, 49–53.

Ibrahim, R. K., and Boulay, B. (1980) 'Purification and some properties of UDP-glucose-O-dihydroxycoumarin 7-O-glucosyltransferase from tobacco cell cultures', *Plant Science Letters*, **19**, 177–184.

James, A. T., and Martin, A. J. P. (1952) 'Gas-liquid partition chromatography: the separation and micro-estimation of volatile fatty acids from formic acid to dodecanoic acid', *Biochem. J.*, **50**, 679–690.

Jordan, L. S., Zurqiyah, A. A., De Mur, A. R., and Clerx, W. A. (1975) 'Metabolism of siduron in Kentucky bluegrass (*Poa pratensis* L.)', *J. Agr. Food Chem.*, **23**, 286–289.

Kamimura, S., Nishikawa, M., Saeki, H., and Takahi, Y. (1974) 'Absorption and metabolism of 3-hydroxy-5-methylisoxazole in plants and the biological activities of its metabolites', *Phytopathology*, **64**, 1273–1281.

Kamsteeg, J., van Brederode, J., and van Nigtevecht, G. (1980). 'The pH-dependent substrate specificity of UDP-glucose; anthocyanidin-3-rhamnosyl glucoside, 5-O-glucosyltransferase in petals of *Silene dioica*: the formation of anthocyanidin-3,5-diglucosides', *Z. Pflanzenphysiol.*, **96**, 87–93.

Kaslander, J., Sijpesteijn, A. K., and van der Kerk, G. J. M. (1961) 'On the transformation of dimethyldithiocarbamate into its β-glucoside by plant tissue', *Biochim. Biophys. Acta*, **52**, 396–397.

Knaak, J. B., Munger, D. M., and McCarthy, J. E. (1970) 'Metabolism of carbofuran in alfalfa and bean plants', *J. Agr. Food Chem.*, **18**, 827–831.

Kobayashi, K., and Ishizuka, K. (1977) 'A mechanism of selectivity of barban. Absorption, translocation and chemical transformation', *J. Pestic. Sci.*, **2**, 59–65.

Kohli, J., Weisgerber, I., and Klein, W. (1976) 'Balance of conversion of [^{14}C]lindane in lettuce in hydroponic culture', *Pest Biochem. Physiol.*, **6**, 91–97.

Krishnan, K., Curbelo, R., Chiha, P., and Noonan, R. C. (1979) 'Design and applications of a high sensitivity gas chromatographic fourier transform infrared system', *J. Chromatog. Sci.*, **17**, 413–416.

Kuhr, R. J., and Casida, J. E. (1967) 'Persistent glycosides of metabolites of methylcarbamate insecticide chemicals formed by hydroxylation in bean plants', *J. Agr. Food Chem.*, **15**, 814–824.

Kunstman, J. L., and Lichtenstein, E. P. (1979) 'Effects of nutrient deficiences in corn plants on the *in vivo* and *in vitro* metabolism of [^{14}C]diazinon', *J. Agr. Food Chem.*, **27**, 770–774.

Laker, M. F. (1980) 'Estimation of sugars and sugar alcohols in biological fluids by gas-liquid chromatography', *J. Chromatog.*, **184**, 457–470.

Lamoureux, G. L., and Frear, D. S. (1979) 'Pesticide metabolism in higher plant: *In vitro* enzyme studies', in *Xenobiotic Metabolism in vitro methods* (G. D. Paulson, D. S. Frear and E. P. Marks, Eds.), American Chemical Society Symposia Series No. 97, pp. 114–128.

Lamoureux, G. L., and Stafford, L. E. (1977) 'Translocation and metabolism of perfluidone (1,1,1-trifluoro-N-[2-methyl-4-(phenylsulfonyl)phenyl]-methanesulfonamide) in peanuts', *J. Agr. Food Chem.*, **25**, 512–517.

Larson, R. L. (1971) 'Glucosylation of quercetin by a maize pollen enzyme', *Phytochem.*, **10**, 3073–3076.

Lawson, M. A., and Russell, G. F. (1980) 'Trace-level analysis of reducing sugars by high-performance liquid chromatography', *J. Food Sci.*, **45**, 1256–1258.

Locke, R. K., Bastone, V. B., and Baron, R. L. (1971) 'Studies of carbamate pesticide metabolism utilizing plant and mammalian cells in culture', *J. Agr. Food Chem.*, **19**, 1205–1209.

Locke, R. K., Chen, J.-Y. T., Damico, J. N., Dusold, L. R., and Sphon, J. A. (1976) 'Identification by physical means of organic moieties of conjugates produced from carbaryl by tobacco cells in suspension culture', *Arch. Env. Cont. Tox.*, **4**, 60–100.

Lubrano, G. J., and Guilbault, G. G. (1978) 'Glucose and L-amino acid electrodes based on enzyme membranes', *Anal. Chim. Acta*, **97**, 229–236.

McLafferty, F. W. (1980) 'Tandem mass spectrometry (ms/ms): a promising new analytical technique for specific component determination in complex mixtures', *Acc. Chem. Res.*, **13**, 33–39.

Mangeot, B. L., Slife, F. E., and Rieck, C. E. (1979) 'Differential metabolism of Metribuzin by two soybean (*Glycine max.*) cultivars', *Weed Sci.*, **27**, 267–269.

Marco, G. J., and Jaworski, E. G. (1964) 'Metabolism of O-phenyl-O'-(4-nitrophenyl)methylphosphonothionate (colep) in plants and animals', *J. Agr. Food Chem.*, **12**, 305–310.

Marquis, L. Y., Shimabakuro, R. H., Stolzenburg, G. E., Feil, V. J., and Zaylskie, R. G. (1979) 'Metabolism and selectivity of fluchloralin in soybean roots', *J. Agr. Food Chem.*, **27**, 1148–1156.

Martinelli, E. M. (1980) 'Gas-liquid chromatographic and mass spectrometric behaviour of plant glycosides, in the form of trimethylsilyl derivatives', *Eur. J. Mass Spectr. Biochem. Med. Environ. Res.*, **1**, 33–43.

Meikle, R. W. (1977) 'Fate of diamidafos (phenyl-N,N'-dimethylphosphordiamidate) in tobacco, cured tobacco and in smoke', *J. Agr. Food Chem.*, **25**, 746–752.

Mellis, J. M., and Kirkwood, R. C. (1980) 'The uptake, translocation and metabolism of Epronaz in selected species', *Pestic. Sci.*, **11**, 324–330.

Mellor, R. B., Roberts, L. M., and Lord, J. M. (1980) 'N-Acetylglucosamine transfer reactions and glycoprotein biosynthesis in castor bean endosperm', *J. Exp. Bot.*, **31**, 993–1003.

Michalezuk, L., and Bandurski, R. S. (1980). 'UDP-glucose: indoleacetic acid glucosyl transferase and indoleacetyl-glucose: myo-inositol indoleacetyl transferase', *Biochem. Biophys. Res. Comm.*, **93**, 588–592.

Mihara, K., and Miyamoto, J. (1974) 'Metabolism of salithion, (2-methoxy-4H-1,3,2-benzodioxaphosphorin-2-sulphide) in rats and plants', *Agr. Biol. Chem.*, **38**, 1913–1924.

Milborrow, B. V. (1970) 'The metabolism of abscisic acid', *J. Exp. Bot.*, **21**, 17–29.

Milborrow, B. V., and Mallaby, R. (1975) 'Occurrence of methyl-(+)-abscisate as an artefact of extraction', *J. Exp. Bot.*, **26**, 741–748.

Miller, J. N., Rocks, B. F., and Burns, D. T. (1977) 'The determination of glucose with immobilized glucose oxidase and peroxidase', *Anal. Chim. Acta*, **93**, 353–356.

Miller, L. P. (1940) 'Formation of β-O-chlorophenyl-gentiobioside in gladiolus corms from absorbed o-chlorophenol', *Contr. Boyce Thompson Inst.*, **11**, 271–279.

More, J. E., Roberts, T. R., and Wright, A. N. (1978) 'Studies of the metabolism of 3-phenoxybenzoic acid in plants', *Pestic. Biochem. Physiol.*, **9**, 268–280.

Morris, H. R. (1980) 'Biomolecular structure determination by mass spectrometry', *Nature*, **286**, 447–452.

Motooka, P. S., Corbin, F. T., and Worsham, A. D. (1977) 'Metabolism of Sandoz 6706 in soybean and sicklepod', *Weed Sci.*, **25**, 9–12.

Müller, W., and Leistner, E. (1978) 'Aglycones and glycosides of oxygenated naphthalenes and a glycosyl transferase from *Juglans*', *Phytochem.*, **17**, 1739–1942.

Mumma, R. O., and Hamilton, R. H. (1979) 'Xenobiotic metabolism in higher plants: *In vitro* tissue and cell culture techniques', in *Xenobiotic Metabolism, in vitro Methods* (G. D. Paulson, D. S. Frear, and E. P. Marks, Eds.), American Chemical Society Symposia Series No. 97, pp. 35–76.

Newsome, W. H. (1980) 'A method for the determination of maleic hydrazide and its β-D-glucoside in foods by high-pressure anion-exchange liquid chromatography', *J. Agr. Food Chem.*, **28**, 270–272.

Niki, Y., Kuwatsuka, S., and Yokomichi, I. (1976) 'Absorption, translocation and metabolism of Chlomethoxynil (X-52) in plants', *Agr. Biol. Chem.*, **40**, 683–690.

Ogawa, K., Tsuda, M., Yamauchi, F., Yamaguchi, I., and Misato, T. (1976) 'Metabolism of 2-sec-butylphenyl-*N*-methylcarbamate (Bassa®, BPMC) in rice plants and its degradation in soils', *J. Pestic. Sci.*, **1**, 219–229.

Ohkawa, H., Nambu, K., and Miyamoto, J. (1980) 'Metabolic fate of fenvalerate (Sumicidin) in bean plants', *J. Pestic. Sci.*, **5**, 215–223.

Österdahl, B-G. (1978) 'Chemical studies on Bryophytes. 19. Application of ^{13}C-NMR in structural elucidation of flavonoid C-glucosides from *Hedwigia ciliata*', *Acta Chem. Scand.*, **B.32**, 93–97.

Ozawa, T., and Takino, Y. (1979) 'Carbon-13 nucelar magnetic resonance spectra of phenolic glycosides isolated from chestnut galls', *Agr. Biol. Chem.*, **43**, 1173–1177.

Percival, F. W., and Bandurski, R. S. (1976) 'Esters of indole-3-acetic acid from *Avena* seeds', *Plant Physiol.*, **58**, 60–67.

Pillmoor, J. B. (1981) *The behaviour of pesticides in plants*. Ph.D. Thesis (Univ. Coll. Nth. Wales, Bangor).

Pillmoor, J. B., Roberts, T. R., and Gaunt, J. K. (1981) 'A study of the stability of wild-oat herbicide conjugates during extraction from plants', *Pestic. Sci.*, (in press).

Pont, V., and Collet, G. F. (1980) Métabolisme du chloro-2-(*p*-chlorophényl)-3-propionate de méthyle et probleme de selectivite', *Phytochem.*, **19**, 1361–1363.

Porter, P. E. (1967) 'Azodrin insecticide' in *Analytical Methods for Pesticides, Plant Growth Regulators and Food Additives* (G. Zweig, Ed.), Academic Press, New York, Vol. V, pp. 193–201, 213–233.

Pree, D. J., and Saunders, J. L. (1974) 'Metabolism of carbofuran in mugho pine', *J. Agr. Food Chem.*, **22**, 620–625.

Quantin, E., Hartmann-Bouillon, M. A., Schuber, F., and Benveniste, P. (1980) 'Latency of uridine diphosphate glucose-sterol-β-D-glucosyltransferase, a plasma membrane-bound enzyme of etiolated maize coleoptiles', *Plant Science Letters*, **17**, 193–199.

Quistad, G. B., Staiger, L. E., and Schooley, D. A. (1978) 'Environmental degradation of the miticide cycloprate. 2. Metabolism by apples and oranges', *J. Agr. Food Chem.*, **26**, 66–70.

Reay, P. F., and Conn, E. E. (1974) 'The purification and properties of a uridine diphosphate glucose: aldehyde cyanohydrin β-glucosyltransferase from sorghum seedlings', *J. Biol. Chem.*, **249**, 5826–5830.

Roberts, T. R. (1978) in *Radiochromatography—The Chromatography and Electrophoresis of Radiolabelled Compounds* (T. R. Roberts, Ed.), Elsevier, Amsterdam, pp. 45–81.

Roberts, T. R., and Wright, A. N. (1981) 'The metabolism of 3-phenoxybenzyl alcohol, a pyrethroid metabolite in plants', *Pestic. Sci.*, **12**, 161–169.

Rouchaud, J., Moons, C., and Meyer, J. A. (1980) 'The metabolism of [^{14}C]Aldicarb in the leaves of sugar beet plants', *Pestic. Sci.*, **11**, 483–492.

Ruzo, L. O., and Casida, J. E. (1979) 'Degradation of Decamethrin on cotton plants', *J. Agr. Food Chem.*, **27**, 572–575.

Sakata, I., and Koshimizu, K. (1979) 'Gas chromatographic optical resolution of DL-menthol and DL-borneol via its acetylglucoside derivatives', *Agr. Biol. Chem.*, **43**, 411–412.

Saleh, M. A., Marei, A. E.-S. M., and Casida, J. E. (1980) 'α-Cyano-3-phenoxybenzyl pyrethroids: derivatisations at the benzylic position', *J. Agr. Food Chem.*, **28**, 592–594.

Sanderman, H., Diesperger, H., and Scheel, D. (1977) 'Metabolism of xenobiotics by plant cell cultures', in *Plant Tissue Culture and its Biotechnological Application: Proceedings of the First International Congress on Medicinal Plant Research, Section B* (W. Barz, E. Reinhard, and M. H. Zenk, Eds.), Springer-Verlag, Berlin, pp. 178–196.

Satô, M., and Hasegawa, M. (1972) 'Transglucosylases in *Cichorium intybus* converting cichoriin to esculin', *Phytochem.*, **11**, 3149–3156.

Scheel, D., and Sandermann, H. (1977) 'Metabolism of DDT and kelthane in cell suspensions of parsley (*Petroselinum hortense*) and soybean (*Glycine max.*)', *Planta*, **133**, 315–320.

Schlepphorst, R., and Barz, W. (1979) 'Metabolism of aromatic acids in plant cell suspension cultures', *Planta Medica*, **36**, 333–342.

Schuphan, I. (1974) 'Metabolism of phenylureas. (3) Metabolism of [^{14}C-methyl]linuron in *Chlorella pyridinosa*', *Chemosphere*, **3**, 131–134.

Shimabukuro, R. H., Lamoureux, G. L., Swanson, H. R., Walsh, W. C., Stafford, L. E., and Frear, D. S. (1973) 'Metabolism of substituted diphenylether herbicides in plants. II. Identification of a new fluorodifen metabolite, *S*-(2-nitro-4-trifluoromethylphenyl)glutathione in peanut', *Pestic. Biochem. Physiol.*, **3**, 483–494.

Shimabukuro, R. H., Walsh, W. C., and Hoerauf, R. A. (1979). 'Metabolism and selectivity of diclofop methyl in wild oat and wheat', *J. Agr. Food Chem.*, **27**, 615–623.

Shimotori, H., and Kuwatsuka, S. (1978) 'Absorption, translocation and metabolism of CNP, a diphenyl ether herbicide, and its amino derivative in rice plants', *J. Pestic. Sci.*, **3**, 267–275.

Silk, W. K., and Jones, R. L. (1975) 'Gibberellin response in lettuce hypocotyl sections', *Plant Physiol.*, **56**, 267–272.

Slife, F. W., Key, J. L., Yamaguchi, S., and Crafts, A. S. (1962) 'Penetration, translocation and metabolism of 2,4-D and 2,4,5-T in wild and cultivated cucumber plants', *Weeds*, **10**, 29–35.

Smith, A. E. (1979) 'The metabolism of 2,4-DB by white clover (*Trifolium repens*) cell suspension cultures', *Weed Sci.*, **27**, 392–396.

Smith, A. E., and Wilkinson, R. E. (1974) 'Differential absorption, translocation and metabolism of metribuzin, [4-amino-6-*tert*-butyl-3-(methylthio)-*as*-triazine-5-[4H]one] by soybean cultivars', *Physiol. Plant.*, **32**, 253–257.

Spitznagle, L. A., Christain, J. E., Ohlrogge, A. J., and Breckinridge, C. E. (1968) 'Study of the absorption, translocation and residue properties of 2,3,5-triiodobenzoic acid in field grown soyabeans', *J. Pharm. Sci.*, **57**, 764–768.

Starr, R. I., and Cunningham, D. J. (1975) 'Degradation of 4-aminopyridine-^{14}C in corn and sorghum plants', *J. Agr. Food Chem.*, **23**, 279–281.

Staver, M. J., Glick, K., and Baisted, D. J. (1978) 'Uridine diphosphate glucose-sterol glucosyltransferase and nucleoside diphosphatase activities in etiolated pea seedlings', *Biochem. J.*, **169**, 297–303.

Stellar, W. A., and Brand, W. W. (1974) 'Analysis of dimethoate-treated grapes for the *N*-hydroxymethyl and de-*N*-methyl metabolites and for their sugar adducts', *J. Agr. Food Chem.*, **22**, 445–449.

Stephenson, G. R., and Ries, S. K. (1969) 'Metabolism of pyrazon in sugar beets and soil', *Weed Sci.*, **17**, 327–331.

Still, G. G. (1968) 'Metabolism of 3,4-dichloropropionanilide in plants: the metabolic fate of the 3,4-dichloroaniline moiety', *Science*, **159**, 992–993.

Still, G. G., and Mansager, E. R. (1972) 'Aryl hydroxylation of isopropyl-3-chlorocarbanilate by soybean plants', *Phytochem.*, **11**, 515–520.

Still, G. G., and Mansager, E. R. (1975) 'High performance liquid and gas-liquid chromatography applied to the purification of pesticide plant and animal metabolites', *Chromatographia*, **8**, 129–135.

Stoddart, J. L., and Jones, R. L. (1977) 'Gibberellin metabolism in excised lettuce hypocotyls: evidence for the formation of gibberellin A$_1$ glucosyl conjugates', *Planta*, **136**, 261–269.

Storm, D. L., and Hassid, W. Z. (1974) 'Partial purification and properties of a β-D-glucosyltransferase occurring in germinating *Phaseolus aureus* seeds', *Plant Physiol.*, **54**, 840–845.

Storm, D. L., and Hassid, W. Z. (1976) 'Occurrence of a D-glucosyltransferase in germinating pea seeds and isolation of an endogenous acceptor', *Biochem. Biophys. Res. Comm.*, **68**, 511–517.

Swanson, C. R., Kadunce, R. E., Hodgson, R. H., and Frear, D. S. (1966) 'Amiben metabolism in plants. I. Isolation and identification of an *N*-glucosyl complex', *Weeds*, **14**, 319–323.

Thornton, J. S., and Stanley, C. W. (1977) 'Gas chromatographic determination of sencor and metabolites in crops and soil', *J. Agr. Food Chem.*, **25**, 380–386.

Tomizawa, C., and Uesugi, Y. (1972) 'Metabolism of *S*-benzyl-*O,O*-diisopropylphosphorothioate (Kitazin P) by mycelial cells of *Pyricularia oryzae*', *Agr. Biol. Chem.*, **36**, 294–300.

Trenck, K. T. v.d., and Sandermann, H. (1978) 'Metabolism of benzo(α)pyrene in cell suspension cultures of parsley and soybean', *Planta*, **141**, 245–251.

Turczan, J. W., Medwick, T., and Plank, W. M. (1978) '220 Hz nuclear magnetic resonance studies of Amygdalin and some related compounds', *J. Assoc. Off. Anal. Chem.*, **61**, 192–207.

Umetsu, N., Fahmy, M. A. H., and Fukuto, T. R. (1979) 'Metabolism of 2,3-dihydro-2,2-dimethyl-7-benzofuranyl(di-n-butylaminosulfenyl)(methyl)-carbamate and 2,3-dihydro-2,2-dimethyl-7-benzofuranyl(morpholinosulfenyl)-(methyl)carbamate in cotton and corn plants', *Pestic. Biochem. Physiol.*, **10**, 104–119.

Verloop, A. (1972) 'Fate of the herbicide dichlobenil in plants', *Residue Reviews*, **43**, 55–70.

Wells, G. B., and Lester, R. L. (1979) 'Rapid separation of acetylated oligosaccharides by reverse-phase high-pressure liquid chromatography', *Anal. Biochem.*, **97**, 184–190.

Williams, E., Meikle, R. W., and Redemann, C. T. (1964) 'Identification of the metabolites of Zectran insecticide in broccoli', *J. Agr. Food Chem.*, **12**, 453–456.

Wing, K. D., and Hammock, B. D. (1979). 'Stereoselectivity of a radioimmunoassay for the insecticide *s*-bioallethrin', *Experientia*, **35**, 1619–1620.

Wright, A. N., and Mathews, B. L. (1976) 'Development of methods for the analysis of crops and soil for residues of benzoylprop-ethyl and its breakdown products', *Pestic. Sci.*, **7**, 339–348.

Wright, A. N., Roberts, T. R., Dutton, A. J., and Doig, M. V. (1980) 'The metabolism of cypermethrin in plants: the conjugation of the cyclopropyl moiety', *Pestic. Biochem. Physiol.*, **13**, 71–80.

Yamaha, T., and Cardini, C. E. (1960) 'The biosynthesis of plant glycosides 1. Monoglucosides', *Arch. Biochem. Biophys.*, **86**, 127–132.

Yih, R. Y., McRae, D. H., and Wilson, H. F. (1968) 'Metabolism of 3',4'-dichloropropionanilide: 3,4-dichloroaniline-lignin complex in rice plants', *Science*, **161**, 376–377.

Zenk, M. H. (1961) '1-(Indole-3-acetyl)-β-D-glucose, a new compound in the metabolism of indole-3-acetic acid in plants', *Nature*, **191**, 493–494.

Zenk, M. H. (1963) 'Isolation, biosynthesis and function of indoleacetic acid conjugates', *Colloq. Intern. Centre Natl. Rech. Sci. (Paris)*, **123**, 241–249.

Zulalian, J., and Blinn, R. C. (1977) 'Metabolism of cytrolane systemic insecticide (mephosfolan), propylene (diethoxyphosphinyl)dithioimidocarbonate in cotton plants', *J. Agr. Food Chem.*, **25**, 1033–1039.

Zurqiyah, A. A., Jordan, L. S., and Jolliffe, V. A. (1976) 'Metabolism of isopropylcarbanilate (Propham) in alfalfa grown in nutrient solution', *Pestic. Biochem. Physiol.*, **6**, 35–45.

Progress in Pesticide Biochemistry, Volume 2
Edited by D. H. Hutson and T. R. Roberts
© 1982 John Wiley & Sons, Ltd

CHAPTER 4

Biochemical mechanisms of amino acid conjugation of xenobiotics

K. R. Huckle and P. Millburn

INTRODUCTION

Aromatic carboxylic acids may undergo conjugation with either a sugar (glucuronic acid in mammals; predominantly glucose in insects and plants) or an amino acid (Millburn, 1978). In mammals, amino acid conjugation is physiologically important for the excretion of: certain endogenous compounds (e.g. bile acids; Haslewood, 1967); dietary constituents and food additives (see

Bridges *et al.*, 1970; James *et al.*, 1972); drugs and their metabolites (see Wan and Riegelman, 1972; Lan *et al.*, 1975), and a variety of other xenobiotics. The conjugation of insecticides in insects has been reviewed by Yang (1976), and peptide bond formation has been reviewed recently by Killenberg and Webster (1980).

Carboxylic acids are often derived from pesticides as a consequence of their metabolism by both animals and plants (viz. by oxidation of alcohols or by hydrolysis of esters, amides and nitriles). For example, DDA (bis(*p*-chlorophenyl)acetic acid, (**1**)), a metabolite of the organochlorine insecticide DDT

(**1**)

(1,1,1-trichloro-2,2-bis(*p*-chlorophenyl)ethane), undergoes amino acid conjugation in a number of plant and animal species (Pinto *et al.*, 1965; Wallcave *et al.*, 1974; Gingell, 1976). 2,4-Dichlorobenzoic acid, a metabolite of the organophosphate insecticide chlorfenvinphos, is conjugated with glycine in some mammals (see Hutson, 1981). Similarly, 3,4-dichlorobenzoylglycine (**3**) is the major urinary metabolite of 3,4-dichlorobenzyl *N*-methyl carbamate (**2**) in

the rat (Knaak and Sullivan, 1968). Photostable, synthetic pyrethroids are a new group of insecticides combining excellent insecticidal activity with low mammalian toxicity. Their metabolism, although complex, gives rise mainly to sugar, sulphate and amino acid conjugates (see Hutson, 1979; Roberts, 1981). Thus, the major degradative pathway of permethrin (**4**) involves hydrolysis of the ester linkage liberating the acidic cyclopropane moiety (**5**) and 3-phenoxybenzylalcohol (**6**), which is subsequently oxidized to 3-phenoxybenzoic acid (**7**). Permethrin metabolism, therefore, results in the production of two metabolites (**5** and **7**), which are both carboxylic acids and can undergo conjugation in mammals (Gaughan *et al.*, 1977, 1978a, b; Ivie and Hunt, 1980), insects

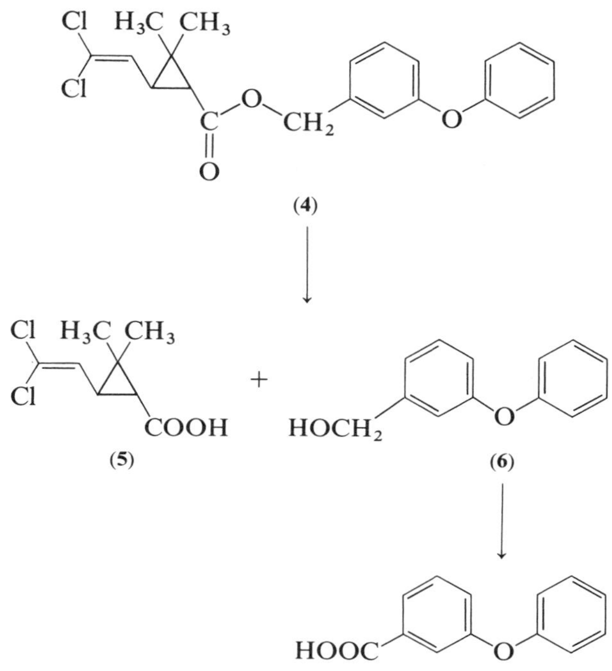

(4)

(5) + (6)

(7)

(Shono *et al.*, 1978), fish (Glickman *et al.*, 1979) and plants (Ohkawa, *et al.*, 1977). Both (5) and (7) have been shown to form mono- and disaccharide conjugates in plants (Wright *et al.*, 1980; More *et al.*, 1978).

Of the pesticidal compounds found in the environment relatively few undergo direct conjugation with amino acids. Probably the most extensively studied examples are the related herbicides 2,4-D (2,4-dichlorophenoxyacetic acid; (8), R = H) and 2,4,5-T (2,4,5-trichlorophenoxyacetic acid; (8), R = Cl). These acids are conjugated with a variety of amino acids in animals (Grunow and Böhme, 1974; Guarino *et al.*, 1977) and in plants (see Mumma and Hamilton, 1976; Arjmand *et al.*, 1978). Indol-3-ylacetic acid (9) and related plant growth modifiers also undergo direct conjugation with amino acids (Andreae and Good, 1961; Bridges *et al.*, 1974; Lethco and Brouwer, 1966).

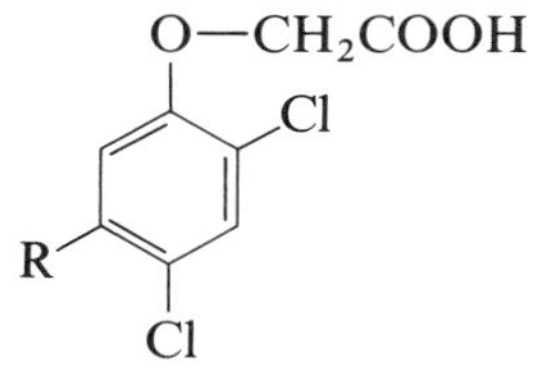

(8)

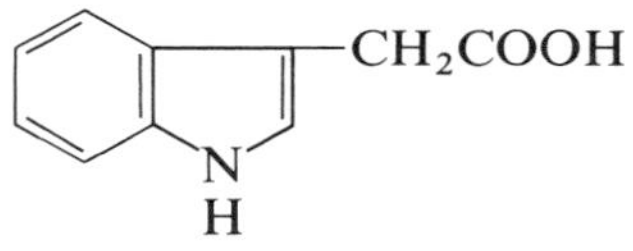

(9)

In addition to the efficaceous agent, other components present in commercial pesticidal formulations may be conjugated with amino acids. Monoalkyl glycol ethers have been widely used as solvents for pesticidal chemicals. Thus, N-isopropoxyacetylglycine (11) is the major urinary metabolite of O-isopropyl-ethanediol [isopropyl oxitol (IPO); (10)] in the rat and dog (Hutson and

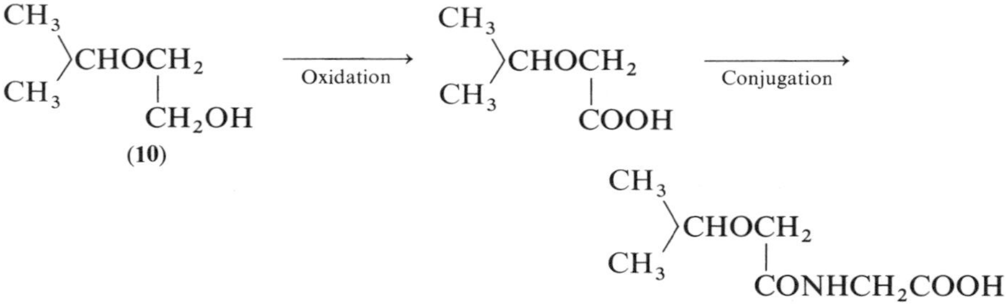

(10) → Oxidation → CH₃CHOCH₂COOH → Conjugation → (11)

Pickering, 1971). The fate of methylenedioxyphenyl synergists has been investigated in animals. Following administration of Tropital (12) to houseflies, five N-piperonyl conjugates were formed, utilizing serine, alanine, glutamic acid, glutamine and glycine (Esaac and Casida, 1968). In mammals, the glycine derivative of piperonylic acid (13) was the major amino acid conjugate detected (Kamienski and Casida, 1970).

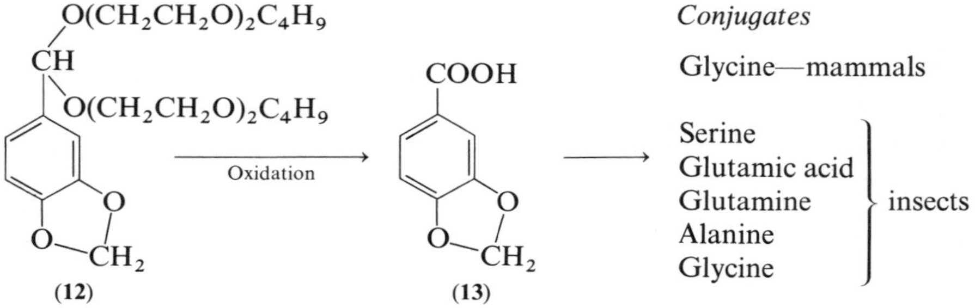

Conjugates

Glycine—mammals

Serine
Glutamic acid
Glutamine   } insects
Alanine
Glycine

(12)   (13)

Species differences are often observed in amino acid conjugation; the amino acid utilized varies with both the animal or plant species and the chemical structure of the participating carboxylic acid (see Millburn, 1978; Climie and Hutson, 1979; Caldwell, 1980). Glycine is the most commonly used amino acid in conjugation, but involvement of a variety of others has been reported (see Hirom *et al.*, 1977; Hirom and Millburn, 1979; Climie and Hutson, 1979). In plants, aspartic acid is most often utilized (see Climie and Hutson, 1979). The recent finding that benzoylaspartate and phenylacetylaspartate are natural constituents of pea seeds (Gianfagna and Davies, 1980) suggests a possible physiological role of this amino acid conjugation in plants. However, the function of the endogenous amino acid conjugates is unknown.

3-Phenoxybenzoic acid (7), a metabolite of some photostable synthetic pyrethroid insecticides (see above), and benzoic acid (14), which occurs naturally in plants (and is, therefore, present in significant amounts in the diet of herbivores and omnivores) and is also widely used as a food preservative, provide examples of this phenomenon (Table 1).

COOH

(14)

With 3-phenoxybenzoic acid, amino acid conjugates involving glycine, glutamic acid or taurine have been reported in non-primate mammals, whereas benzoic acid is eliminated principally as its glycine conjugate (hippuric acid) in these species (Huckle *et al.*, 1981b). In the mallard duck, 3-phenoxybenzoic acid is excreted mainly as the glycylvaline dipeptide conjugate (15) (Huckle *et al.*,

$$CH(CH_3)_2$$
CONHCH$_2$CONHCHCOOH
(15)

1981a). This is the first report of the involvement of valine in the metabolism of a xenobiotic in animals, although 2,4-D (8, R = H) is metabolized to a valine conjugate by soya bean cotyledon callus tissue (Feung *et al.*, 1973). Although rarely encountered (see Hirom and Millburn, 1981), dipeptide conjugation has been reported previously in pesticide metabolism. A dipeptide containing DDA (1), aspartic acid and serine is excreted by rats administered DDT (Pinto *et al.*, 1965).

Such remarkable species diversity in amino acid conjugation makes it very difficult to predict the pattern of metabolism of aromatic acids such as 3-phenoxybenzoic acid in man and serves further to illustrate how these differences can complicate the safety evaluation procedure (see Hutson, 1981). Much of the *in vivo* information has been collected using only a small number of model compounds and this has led to the occurrence in the literature of some equivocal generalizations between species and amino acid used for conjugation (see Climie and Hutson, 1979). By contrast, comparatively little work has been performed on the elucidation of the biochemistry behind such differences, although the basic enzymology of the reaction has been known for some time. It is the authors' belief that further predictive information may be obtained by the investigation of the biochemical mechanisms of amino acid conjugation.

Table 1   Amino acid conjugation of benzoic acid and 3-phenoxybenzoic acid

Species	Major conjugate formed with:	
	Benzoic acid	3-Phenoxybenzoic acid
Vertebrates		
1. Mammals		
(i) Placental mammals		
Primates	Gly	—
Non-primate mammals	Gly (Glu*)	Gly or Glu or Tau†
(ii) Marsupials (opossum)	Gly	—
2. Birds		
(i) Galliformes (hens, turkeys) and Anseriformes (ducks, geese)	Orn	Gly–Val‡
(ii) Columbiformes (pigeons)	Gly	—
(iii) Passeriformes (crows) and Psittaciformes (parrots)	n.d.	—
3. Reptiles		
(snakes, lizards, tortoises, gecko, turtles)	Orn	—
4. Amphibia		
(frogs)	Gly	—
5. Fish		
(dogfish shark, goose fish, flounder)	Gly§	—
Invertebrates		
1. Insects		
(locusts, mosquitoes, houseflies)	Gly	Gly
2. Crustacea		
(terrestrial species: sea slater, wood louse)	Gly§	—
(land crab, freshwater crayfish)	n.d.§	—
3. Arachnids		
(spiders, ticks, harvestmen, scorpions)	Arg¶	—
4. Myriapods		
(millipedes)	Arg§	—
5. Onychophora		
(*Peripatus*)	His	—
Plants		
1. Higher plants	Asp	—

Benzoic acid data taken from Millburn (1978) and references therein
3-Phenoxybenzoic acid data from Huckle *et al.*, (1981a,b)
* Benzoylglutamic acid is a minor metabolite in certain species of bats
† Major amino acid conjugate varies with species of mammal
‡ Dipeptide conjugate, in the mallard duck (see text)
§ Data for amino and nitrobenzoic acids
¶ Secondary products are formed by further metabolism; benzoylarginine can be decarboxylated to give benzoylagmatine, and can also be converted into benzoylornithine, benzoylcitrulline and benzoylglutamic acid
Key: Gly = glycine, Glu = glutamic acid, Tau = taurine, Orn = ornithine, Arg = arginine, His = histidine, Asp = aspartic acid, n.d. = no detectable amino acid conjugates.

The information currently available has been derived almost exclusively from mammalian studies; there is a paucity of data concerning the mechanisms of amino acid conjugation in insects or plants. Current knowledge on these mechanisms in mammals is reviewed in this chapter.

HISTORICAL ASPECTS

Hippurate (benzoylglycine) synthesis in man was first established in 1842 by Keller (see Conti and Bickel, 1977). Quick (1931, 1932a, b) later showed that the formation of this amino acid conjugate and several of its congeners occurred mainly in the kidney of the dog, but in both liver and kidney tissue from man and rabbit. Kielly and Schneider (1950) and Leuthardt and Nielsen (1951; cited in Irjala, 1972) found that the hippuric acid synthesizing enzymes were located predominantly in the mitochondrial fractions of liver and kidney.

Borsook and Dubnoff (1947) showed that energy was required from oxidative reactions for the synthesis of hippurate in rat liver homogenates. This energy is provided by the oxidative reactions of the citric acid cycle or, under anaerobic conditions, from ATP (Cohen and McGilvery, 1946, 1947a, b). The involvement of coenzyme A (CoA) in the reaction was first reported by Chantrenne (1951), who additionally postulated the formation of an energy-rich intermediate or 'CoA compound'. Benzoyl phosphate (see Chantrenne, 1951), N-phospho-glycine (Cohen and McGilvery, 1947a) and S-benzoyl-CoA (Cohen and Ochoa, 1951, cited in Schachter and Taggart, 1953) were all proposed, the CoA derivative being confirmed as the activated intermediate in the enzymic synthesis of hippurate by a soluble pig kidney preparation (Schachter and Taggart, 1953). These workers further proposed the general reaction mechanism (Figure 1) presently accepted for the renal and hepatic conjugation of other exogenous aromatic carboxylic acids in mammals (Moldave and Meister, 1957; Jones, 1959; Forman et al., 1971).

Amino acid conjugation occurs with the formation of an amide bond. The first step of the reaction occurs with the interaction of the aromatic carboxylic acid with CoA to form an acyl-CoA thioester (Figure 1a). This appears to be catalysed by medium-chain fatty acyl-CoA synthetase (EC 6.2.1.2) (Mahler et al., 1953). Conjugation of the activated xenobiotic with the amino acid takes place in the second step (Figure 1b) and is catalysed by an acyl-CoA: amino

$$RCOOH + CoASH + ATP \xrightarrow{Mg^{2+}} RCOSCoA + AMP + PP_i$$

(a)

$$RCOSCoA + H_2NCH_2COOH \longrightarrow RCONHCH_2COOH + CoASH$$

(b)

Figure 1   Enzymatic conjugation of aromatic carboxylic acids (RCOOH) with glycine

acid N-acyltransferase (e.g. glycine N-acyltransferase; EC 2.3.1.13). Thus, the formation of amino acid conjugates may be contrasted with the mechanism of most other conjugation reactions (e.g. glucuronidation, sulphation, acetylation and methylation). In amino acid conjugation, it is the xenobiotic rather than the endogenous conjugating moiety which is activated prior to the transferase action.

Some amino acid conjugates, which do not involve linkage through the α-amino group have been reported. These include alanine conjugates of N,N-dialkylthiocarbamates (Kaslander et al., 1962) and alanine, glycine or serine conjugates of 3-amino-1,2,3-triazole (Carter, 1965). The mechanism(s) of these conjugations is unknown.

ACYL-CoA SYNTHETASES

The enzymes involved in the initial stage of the β-oxidation of fatty acids, the fatty acyl-CoA synthetases, catalyse the esterification of these and some aromatic carboxylic acids with CoA. They are classified into four groups based on specificity: short-chain (acetyl-CoA synthetase; acetate: CoA ligase (AMP); EC 6.2.1.1), medium-chain (butyryl-CoA synthetase; medium-chain fatty acid: CoA ligase (AMP); EC 6.2.1.2) and long-chain fatty acyl-CoA synthetase (acyl-CoA synthetase; long-chain fatty acid: CoA ligase (AMP); EC 6.2.1.3), which are all ATP-dependent and follow the reaction shown in Figure 2a, and the GTP-dependent enzyme (medium-long-chain fatty acid: CoA ligase (GDP) (EC 6.2.1.10), which follows the reaction shown in Figure 2b.

The properties of these enzymes have been reviewed (Londesborough and Webster, 1974; Groot et al., 1976; Park, 1978). Medium-chain fatty acyl-CoA synthetase appears to be responsible for the activation of several exogenous aromatic carboxylic acids, and is therefore considered in more detail.

Medium-chain fatty acyl-CoA synthetase (EC 6.2.1.2.)

This enzyme was partially purified from beef liver mitochondria by Mahler et al. (1953) and has since been reported in a large variety of mammalian tissues, higher plants, yeasts and bacteria (see Jenks, 1962; Londesborough and Webster, 1974).

$$RCOOH + ATP + CoASH \underset{(a)}{\overset{Mg^{2+}}{\rightleftharpoons}} RCOSCoA + AMP + PP_i$$

$$RCOOH + GTP + CoASH \underset{(b)}{\overset{Mg^{2+}}{\rightleftharpoons}} RCOSCoA + GDP + P_i$$

Figure 2

Substrate specificity

Mahler's enzyme showed broad substrate specificity, being active with a variety of aliphatic (C_4–C_{12}; C_7 optimal) and aromatic carboxylic acids (including benzoic acid, phenylacetic acid and 2,4-D). Schachter and Taggart (1954) obtained a partially purified fraction from beef liver mitochondria which catalysed the formation of CoA thioesters of several *para*- and *meta*-substituted benzoates and heterocyclic acids, but did not activate the *ortho*-hydroxy acids salicylate and homogentisate. Killenberg *et al.* (1971) showed that *ortho*-methoxybenzoate and *ortho*-aminobenzoate (anthranilate) could serve as substrates for this enzyme, but confirmed that salicylate (and *para*-amino-salicylate) could not. Moreover, they described the isolation of a 'salicylate': CoA ligase (AMP), which could be distinguished from Mahler's enzyme when partially purified from the same source. The former was active with hexanoate, benzoate, *ortho*-methoxybenzoate and anthranilate, as well as with salicylate and *para*-aminosalicylate and required ATP and Mg^{2+} (Killenberg *et al.*, 1971). Groot and Scheek (1976) have isolated three soluble ATP-dependent acyl-CoA synthetases from guinea pig liver mitochondria: a medium-chain acyl-CoA synthetase, a salicylate-activating enzyme, and a propionyl-CoA synthetase accepting acetate, propionate and butyrate, but with a high preference for propionate.

The substrate specificity of medium-chain fatty acyl-CoA synthetases can vary with the tissue of origin. Thus, an enzyme has been purified from beef heart mitochondria which, unlike related enzymes in beef liver mitochondria (Mahler *et al.*, 1953; Killenberg *et al.*, 1971), does not activate aromatic fatty acids or aliphatic acids greater than C_7 (Webster *et al.*, 1965). It would appear, however, that no acyl-CoA synthetases have been isolated to date which exclusively activate aromatic carboxylic acids. Plants may contain such enzymes, but their substrate specificities remain to be determined (Walton and Butt, 1971; Hahlbrock and Grisebach 1970; Renjeva *et al.*, 1976).

Subcellular location

Several schemes have been proposed in attempts to understand the manner in which acyl-CoA synthesizing enzymes are arranged in mitochondria (see Garland *et al.*, 1970). There appear, however, to be some conflicting reports in the literature with respect to enzyme location and specific activities. These discrepancies may have resulted from the different methods used in isolating the sub-mitochondrial fractions, and in assaying the enzymic activities (see Lippel and Beattie, 1970). Aas and Bremer (1968) were the first to demonstrate that medium-chain fatty acyl-CoA synthetase is located in the matrix of rat liver mitochondria. Their assay, based on the enzymic transfer of acyl groups from acyl-CoA to carnitine ('carnitine-trapping'), utilized pre-swollen sonicated mitochondria. Their finding has subsequently been corroborated using different assay techniques (Garland *et al.*, 1970; Gatley and Sherratt, 1976; Groot *et al.*, 1974).

The situation may, however, be complicated by the fact that several different medium-chain fatty acyl-CoA synthetases may occur in the same subfraction of a mammalian tissue. The butyryl-CoA synthetase activity from beef liver mitochondria has been resolved into two inter-convertible subfractions, which may, however, be derived from a single protein (Bar-Tana and Rose, 1968a,b; Bar-Tana *et al.*, 1968). Similarly, crude extracts of dog kidney mitochondria have both butyryl-CoA and benzoyl-CoA synthetase activities, but a fraction which activates butyrate, but not benzoate, can be obtained from them (see Londesborough and Webster, 1974).

Molecular properties

Medium-chain fatty acyl-CoA synthetase from rat liver mitochondria has a molecular weight of 47,000 following gel chromatography on Sephadex G-100 (Groot *et al.*, 1974). The enzyme from beef liver mitochondria was found to have a molecular weight of 65,000 by thin-layer gel chromatography and SDS polyacrylamide-gel electrophoresis (Osmundsen and Park, 1975). More recently, a value of approximately 40,000 has been reported for this enzyme when it was isolated as an acetone-dried powder (Johnston *et al.*, 1979). Sedimentation equilibrium analysis suggested the presence in solution of higher molecular weight forms of the enzyme. These could also be obtained by extracting the enzyme from the mitochondrial powder in non-reducing conditions. The enzyme appears to be part of a complex, the 40,000 molecular weight protein being liberated by the acetone treatment of the mitochondria. The enzyme is inhibited by sulphydryl reagents [*p*-chloromercuribenzoate, 5,5'-dithiobis (2-nitrobenzoic acid) (DTNB) and iodoacetamide] and has at least one available thiol group per molecule. Furthermore, a non-linear relationship between enzymic activity and protein concentration suggests an equilibrium between inactive monomer and active dimer. A five to six-fold activation by bovine serum albumin was observed, and since free thiol groups on the albumin molecule were necessary for activation, Johnston *et al.* (1979) concluded that association of the enzyme monomer with albumin, involving the formation of disulphide bonds, results in an activation analogous to that obtained on dimer formation.

Catalytic properties

Some conflicting reports have been published on the mechanism of action of medium-chain fatty acyl-CoA synthetase. The mechanism originally proposed for acetyl-CoA synthetase from yeast (Berg, 1956), and often extended to the other acyl-CoA synthetases, is a two-step process (Figure 3) in which the formation of an enzyme-bound acyladenylate intermediate is postulated. Kellerman (1958) suggested that the activation of benzoic acid proceeded via the

$$RCOOH + ATP \rightleftharpoons [RCOAMP] + PP_i$$

(a)

$$[RCOAMP] + CoASH \rightleftharpoons [RCOSCoA] + AMP$$

(b)

Figure 3

intermediate benzoyladenylate, but was unable to demonstrate its formation by the liberation of pyrophosphate from ATP. He concluded that benzoyladenylate remained tightly bound to the enzyme during catalysis. Moldave and Meister (1957) demonstrated that phenylacetyladenylate, in the presence of added CoA, could replace phenylacetate and ATP in the synthesis of phenylacetylglutamine in a human liver mitochondrial system, supporting this hypothesis. Furthermore, enzyme-bound acetyladenylate and butyryladenylate can be made from ATP, acetate or butyrate, and large amounts of the respective activating enzyme under appropriate conditions (Webster, 1963; Webster and Campagnari, 1962). However, while acyladenylates are good substrates for most, if not all, fatty acyl-CoA synthetases, it is far from clear whether they are obligatory intermediates in the over-all reaction.

Jones *et al.*, (1953) had earlier suggested a ping-pong mechanism. This did not involve an acyl-AMP intermediate on the basis of isotope exchange studies. Further support for this alternative mechanism has come from product inhibition studies (Graham and Park, 1969). Bar-Tana *et al.* (1968) separated the beef liver enzyme into two inter-convertible subfractions, one apparently catalysing the reaction by an ordered mechanism (Bar-Tana and Rose, 1968b), the other (Bar-Tana and Rose, 1968a) by the Berg mechanism. Recently, studies with acetyl-CoA synthetase from several sources have produced somewhat equivocal evidence for the acyladenylate intermediate (see Park, 1978).

To comply with the Berg mechanism, pyrophosphate must dissociate from the enzyme immediately before CoA reacts in the forward reaction (Cleland, 1963). For the medium-chain enzyme, inhibition by pyrophosphate is not, however, of the simple competitive type. Other observations (see Londesborough and Webster, 1974) show that, for certain preparations of medium-chain acyl-CoA synthetase, pyrophosphate does not dissociate immediately before addition of CoA, and therefore a quaternary enzyme-substrate complex may occur. This opens up the possibility for concerted mechanisms, although the transitory occurrence of acyladenylates in the central enzyme-substrate complexes cannot be excluded by initial rate data. A satisfactory mechanism has not been convincingly established for this fatty acyl-CoA synthetase, and the information presently available suggests there may be important mechanistic differences between the members of this group of enzymes. The kinetic properties of related enzyme activities (short- and long-chain fatty acyl-CoA synthetases) have recently been reported from rat liver mitochondria (Farrar and Plowman, 1979; Phillip and Parsons, 1979a, b).

ACYL-CoA: AMINO ACID N-ACYLTRANSFERASES

Schachter and Taggart (1953) first discovered acyl-CoA: amino acid N-acyltransferase activity in preparations of pig kidney cortex. Later, they showed that a partially purified enzyme from beef liver mitochondria catalysed acyl transfer from a wide variety of aliphatic and aromatic acyl-CoA substrates specifically to glycine (Schachter and Taggart, 1954). This step is catalysed by glycine N-acyltransferase (Gly. N.A.T., glycine N-acylase; EC 2.3.1.13). Subsequent investigations (see Table 2) have revealed much concerning the distribution, kinetic and molecular properties of this enzyme(s); these are discussed below. The transfer of acyl-CoA substrates to glutamine (Moldave and Meister, 1957; Webster *et al.*, 1976) and taurine (James, 1978) has also been described (see Table 2).

Acyl donor substrate specificity

Schachter and Taggart's (1954) enzyme preparation from bovine liver mitochondria exhibited broad substrate specificity; activity was reported toward a variety of aliphatic acyl-CoA derivatives (acetyl, propionyl and valeryl) in addition to benzoyl-CoA. No activity was found, however, with succinyl-CoA. Forman *et al.* (1971) demonstrated that Gly. N.A.T. activity, purified from a bovine liver mitochondrial preparation, could utilize both salicyl-CoA and benzoyl-CoA as substrates. Bartlett and Gompertz (1974), in a study of the substrate specificity of bovine liver Gly. N.A.T. and acylglycine excretion in the organic acidaemias, found that isovaleryl-CoA (**16**), a metabolite of leucine, was the most active of the saturated acyl-CoA thioesters used (activity, in descending order: isovaleryl-CoA, 2-methylbutyryl-CoA (**17**), propionyl-CoA, acetyl-CoA), having an activity similar to that of benzoyl-CoA. The unsaturated derivative, tiglyl-CoA (*trans*-2-methyl-2-butenoyl-CoA; **18**), which like (**17**) is an intermediate in the catabolism of isoleucine, was, however, the most active substrate, having a reactivity three times that of benzoyl-CoA. No reactivity was observed with methylmalonyl-CoA (**19**).

$$CH_3 \atop CH_3 \Large\rangle CHCH_2COSCoA$$

(16)

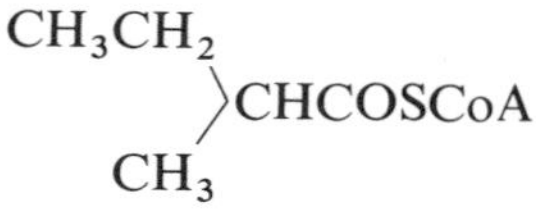

$$CH_3CH_2 \atop CH_3 \Large\rangle CHCOSCoA$$

(17)

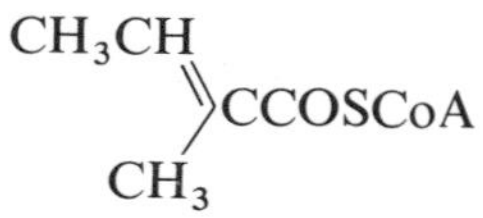

$$CH_3CH \atop CH_3 \Large\rangle CCOSCoA$$

(18)

$$HOOC \atop CH_3 \Large\rangle CHCOSCoA$$

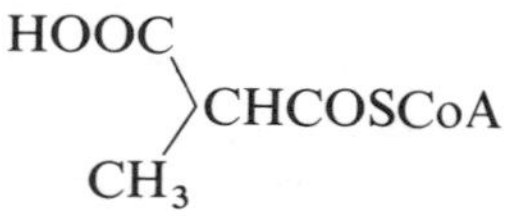

(19)

Recently, using acyl-CoA: amino acid N-acyltransferases purified to near homogeneity (Webster *et al.*, 1976; Nandi *et al.*, 1979), more information has become available on the acyl group specificity of these enzymes. Moldave and Meister (1957) had reported acylation of both glutamine and glycine with phenylacetyl-CoA, but not indol-3-ylacetyl-CoA, as the acyl donor; the same preparations, from human liver and kidney, catalysed the formation of hippurate from benzoyl-CoA and glycine. With liver mitochondrial enzymes from Rhesus monkey and man, Webster and coworkers could not, however, demonstrate acylation of glycine with either phenylacetyl-CoA or indol-3-ylacetyl-CoA, but were able to show acyl transfer to glutamine with both arylacetyl-CoA donors. Their data suggested the presence of two enzymes, i.e. glycine and L-glutamine specific N-acyltransferases, each with differing acyl group specificity. Benzoyl-, salicyl- and, to a lesser extent, butyryl-CoA served as acyl donors for the glycine enzyme, whereas the glutamine enzyme utilized phenylacetyl-CoA or indol-3-ylacetyl-CoA. Moreover, with the monkey liver enzymes, acyl-CoA substrates for one transferase inhibited the other.

The most convincing evidence to date, in support of the existence of a family of closely related transferases with differing acyl group specificity, is that of Nandi *et al.* (1979). These workers describe the isolation and purification, to near homogeneity, of two distinct acyl-CoA: amino acid N-acyltransferases from bovine liver mitochondria. One transferase principally utilizes benzoyl-CoA ('*benzoyltransferase*') while the other uses phenylacetyl-CoA ('*phenylacetyltransferase*'). Benzoyltransferase can be separated from phenylacetyltransferase by chromatography of mitochondrial preparations on hydroxyapatite columns. Previously, both activities had been ascribed to a single protein. Salicyl-CoA and certain short-chain and branched-chain fatty acyl-CoA esters (acetyl, propionyl, butyryl, isobutyryl, isovaleryl, 3-methylcrotonyl and tiglyl) additionally serve as substrates for benzoyltransferase, while the other enzyme specifically uses phenylacetyl-CoA or indol-3-ylacetyl-CoA. Malonyl-CoA does not serve as a substrate with benzoyltransferase, and only marginal activity was observed with methylmalonyl-CoA. Acyl-CoA substrates for one transferase are not substrates for the other enzyme, but competitively inhibit it with respect to the preferred acyl-CoA substrate.

Acyl acceptor substrate specificity

The partially purified enzyme activity originally isolated from bovine liver mitochondria (Schachter and Taggart, 1954) was specific for glycine. None of the other natural α-amino-acids tested would substitute for glycine as the acyl acceptor. Dipeptides (e.g. glycylglycine, glycylalanine, glycylleucine), tripeptides (glycylglycylglycine), N-substituted glycines (sarcosine, acetylglycine) and several other compounds were also inactive as acyl acceptors. A rat liver enzyme preparation utilized only glycine for the conjugation of phenylacetic acid; no

Table 2 Properties of acyl-CoA : amino acid N-acyltransferases

Enzyme source	Enzyme fold purification	Substrate specificity		Optimum pH (range)	Kinetic properties			Molecular weight	Reference
		Preferred amino acid acceptors	Preferred aromatic acyl-CoA		Apparent K_m variable substrate conc (mM)	Fixed substrate conc (mM)	v_{max} μ moles conj min/mg protein		
Bovine liver mitochondria	53	Gly	Benzoyl	8.2 (7.3–9.7)	14 Gly 0.31 Benz	0.0002 Benz 0.06 Gly	1.2 × 10^{-4}*	190000 by ultracentrifugation	Schachter and Taggart (1954)
Rat liver mitochondria	782	Gly	Benzoyl	8.7 (8.1–9.2)	—	—	—	—	Brandt et al. (1968)
Human liver and kidney mitochondria	87 (liver) 49 (kidney)	Gln	Phenylacetyl	— (9.0–10.0)	—	—	—	—	Moldave and Meister (1957)
Rat kidney mitochondria	—	Gly	Nicotinyl	—	—	—	—	—	Jones (1959)
Bovine and human liver mitochondria	6	Gly	Benzoyl Salicyl	—	—	—	—	—	Tishler and Goldman (1970)
Bovine liver mitochondria	—	Gly	Salicyl Benzoyl	—	20 Gly 10 Gly 0.008 Sal 0.015 Benz	0.064 Sal 0.064 Benz 150 Gly 80 Gly	—	32000 on Sephadex G-100; 4 protein peaks on disc gel electrophoresis	Forman et al. (1971)
Bovine liver mitochondria	46	Gly	Benzoyl	—	0.009 Benz 3.3 Gly	0.06† Gly	10.4 Gly	—	Bartlett and Gompertz (1974)

Source	Specific activity*	Amino acid	Acyl substrate	pH	K_m and substrate	K_m and substrate	K_m and substrate	Molecular weight/comments	Reference
Rhesus monkey liver mitochondria	30 11	Gly Gln	Benzoyl Phenylacetyl	—	20 Gly 0.006 Benz 600 Gln 0.035 Phen	0.06 Benz 100 Gly 0.15 Phen 150 Gln	—	24000 for Gly enzyme	Webster et al. (1976)
Human liver mitochondria	23 9	Gly Gln	Benzoyl Phenylacetyl	—	—	—	—	—	Webster et al. (1976)
Bovine liver mitochondria	15	Gly	p-Azidobenzoyl Benzoyl	—	0.026 p Benz 0.037 Benz	100 Gly 100 Gly	0.012‡ Gly 0.0088 Gly	35000, single polypeptide chain, one active site per enzyme molecule	Lau et al. (1977)
Rat and rabbit liver and kidney mitochondria	—	Gly	Phenylacetyl	9.3 (8.0–10.0)	20 Gly 150 Phen	2 Phen 400 Gly	0.083§ Phen 0.084 Gly	—	James and Bend (1978a)
Stingray kidney mitochondria	—	Tau	Phenylacetyl	(7.7–8.2)	0.2 Phen	100 Tau	—	—	James (1978)
Bovine liver mitochondria	24 940	Gly	Benzoyl ⎱ Phenylacetyl ⎰	(8.4–8.6)	3 Gly 0.020 Benz 20 Gly 0.020 Phen	0.1 Benz 200 Gly 100 Phen 100 Gly	—	33000 both enzymes, determined by various methods (see text)	Nandi et al. (1979)
Bovine liver mitochondria	—	Gly	Benzoyl	—	5 Gly 0.025 Benz	0.1 Benz 20 Gly	—	—	Gatley (1979)

* Units are: mmol/min per 100 µg protein
† Tiglyl-CoA at non-specified concentration was fixed substrate
‡ Units are: µmole-SH released/min
§ Data for Triton-solubilized renal preparation from rabbit
Key: Gly = glycine, Gln = glutamine, Tau = taurine, Benz = benzoate, Phen = phenylacetate, Sal = salicylate, p Benz = p-azidobenzoate

phenylacetylglutamine was detected (Moldave and Meister, 1957). The benzoyl and phenylacetyl Gly.N.A.T. activities purified recently from cow liver (Nandi *et al.*, 1979) do not show the same exclusivity toward glycine as an acyl acceptor, although this is the preferred amino acid substrate. Benzoyltransferase showed 10% to 20% of the rate observed with glycine toward L-asparagine, when assayed at similar concentrations (0.1 M); the corresponding rate with L-glutamine was 5% to 7%. The apparent K_m values for these weak alternative acyl acceptors were appreciably higher (> 100 mM) than that of ~3 mM found for glycine. Phenylacetyltransferase also had higher apparent K_m values for L-asparagine and L-glutamine (> 100 mM) than for glycine (20 mM). Other L-amino acids, as well as D-alanine, D-glutamine, methylamine, *N,N*-dimethylglycine and aminomethylphosphonate, were not acceptors, the last three compounds acted as weak inhibitors of benzoyltransferase with respect to glycine.

The observation that L-glutamine can serve as a weak acyl acceptor for bovine liver phenylacetyltransferase is significant as regards species differences in amino acid conjugation. An investigation by James *et al.* (1972) into the species variations in phenylacetic acid metabolism concluded that glutamine conjugation in mammals is restricted to species belonging to anthropoid families (i.e. monkeys, apes and man). However, it has been shown recently (see Hirom and Millburn, 1979) that glutamine conjugates of arylacetic acids can be synthesized by certain non-primate mammals (e.g. 2-naphthylacetylglutamine in rat, rabbit and ferret). If, like the cow, other non-primate mammals possess a phenylacetyltransferase for which L-glutamine is a weak amino acid substrate, then this would account for the glutamine conjugation found *in vivo*. The glutamine acceptor activity of phenylacetyltransferase certainly offers an explanation for the traces of phenylacetylglutamine found in cows' milk by Schwartz and Pallansch (1962).

Partially purified fractions from human liver and kidney mitochondria (Moldave and Meister, 1957) catalysed the condensation of phenylacetyl-CoA with either glutamine or glycine, although with the kidney preparation, a seven-fold greater activity toward the former amino acid was seen. Glutamate, phenylalanine or leucine, however, were inactive as acyl acceptors in this system. Two separate acyl-CoA: amino-acid *N*-acyltransferases are now known to be present in liver mitochondria from Rhesus monkey and man, one specific for glycine and benzoyl-CoA, and the other for L-glutamine and phenylacetyl-CoA. In each case, no acceptor activity was noted with other amino acids or nitrogenous compounds investigated as acyl acceptors (i.e. 18 naturally occurring amino acids, D-alanine and D-glutamine, taurine and ammonium chloride). Comparing the human and monkey enzymes with those from the cow, it appears that the pattern of acyl-CoA specificity was established for benzoyltransferase and phenylacetyltransferase before the evolutionary divergence of primates from ungulates, whereas the amino acid specificity of the latter enzyme changed in primate evolution from glycine (found in prosimians and non-primate mammals) to L-glutamine. No information is available concerning the acyl acceptor specificity of the taurine N-acyltransferase activity reported by James (1978).

Tissue location

Renal and hepatic tissues are the principal sites where acyl-CoA: amino acid N-acyltransferase activities have been most extensively studied (see Table 2). James and Bend (1978a) showed that, in rat and rabbit, hepatic activities of Gly.N.A.T. were lower than those of kidney. This is in general agreement with reports where the over-all conjugation reaction was measured (see Quick, 1932a). Kao et al. (1978) have demonstrated recently, using isolated hepatocyte and renal tubule cell preparations, that hippurate formation in carnivores (ferret and dog) occurs predominantly in the kidney rather than the liver.

No measurable Gly.N.A.T. activity was detected in rabbit lung tissue (James and Bend, 1978a), and extracts of pigeon or chicken liver had a barely detectable activity when tested with either glycine or ornithine (Schachter and Taggart, 1954). Rabbit placental mitochondria had high acyl-CoA hydrolase activity, but no conjugating activity, at least from gestation age 17 days, suggesting that it is not a site for amino acid conjugation. Similarly, transferase activity in suspensions of rabbit small intestine were very low (James and Bend, 1978b). Strahl and Barr (1971) also demonstrated a low rate of glycine conjugation, using ^{14}C-labelled benzoic acid, in the tissue compartment of rat intestinal slices and everted intestinal preparations, and Shirkey et al., (1979), using isolated cells from rat small intestine, likewise found poor amino acid conjugation. Cohen and McGilvery (1946) could not demonstrate synthesis of p-amino-hippuric acid in rat tissue slices from brain, spleen, skeletal muscle, heart and testicles. No information is available concerning the relative tissue distribution of taurine and glutamine N-acyltransferase activities.

Subcellular location

Gly.N.A.T. was first shown to be a mitochondrial enzyme by Schachter and Taggart (1954). Subsequently, it has been located in the matrix by experiments with mitochondria treated with digitonin to remove the outer membrane and the contents of the inter-membrane space (Gatley and Sherratt, 1976). Under such conditions, hippurate synthesis is still observed. In rat kidney fractions, James and Bend (1978a) showed that Gly.N.A.T. activity paralleled that of glutamate dehydrogenase, which is widely accepted as a marker enzyme for the mitochondrial matrix (De Duve et al., 1955; Sottocasa et al., 1967). Further, the effect of several different agents that disrupt mitochondrial membranes showed latency of Gly.N.A.T. activity in rat and rabbit kidney preparations. Sonication of mitochondrial suspensions or their treatment with a membrane perturbant (Triton X-100) gave the most marked increases in activity. James and Bend, therefore, inferred that Gly.N.A.T. is located in the mitochondrial matrix and is not membrane-bound. Similar conclusions were made for taurine N-acyltransferase from stingray mitochondria (James, 1978) and for glutamine conjugating activity in Rhesus monkey and human liver (Webster *et al.*, 1976).

Molecular properties

Schachter and Taggart (1954) reported a molecular weight for partially purified bovine liver Gly.N.A.T. of 190,000, measured by analytical centrifugation. By contrast, gel filtration (Sephadex G-100) of a similar preparation gave a molecular weight of approximately 32,000 (Forman *et al.*, 1971). Recent studies by Lau *et al.* (1977) and Nandi *et al.* (1979) support those of Forman and colleagues. Lau *et al.* found a molecular weight in the range 34,000 (SDS polyacrylamide-gel electrophoresis) to 36,000 (gel filtration on Sephadex G-100) for Gly.N.A.T. from a highly purified bovine liver mitochondrial preparation. Nandi *et al.* have recently confirmed these findings following purification of two bovine liver acyl-CoA: amino acid N-acyltransferases to near homogeneity. The purification procedures for benzoyltransferase and phenylacetyltransferase were quite similar: ammonium sulphate precipitation, aluminium hydroxide gel solution, followed by gel chromatography on Bio-Gel P100 followed by P60 or Sephadex G-100. Benzoyltransferase was then separated from phenylacetyltransferase by chromatography on hydroxyapatite columns. The ratio of benzoyltransferase to phenylacetyltransferase was 19:1 in the crude mitochondrial preparation. Both enzymes had molecular weights near 33,000, as estimated by SDS disc-gel electrophoresis, gel filtration (Sephadex G-100) and sucrose density centrifugation. Benzoyltransferase appeared as a single band on SDS disc-gel electrophoresis, whereas three protein bands, each with enzymic activity, were found on conventional polyacrylamide gel electrophoresis in the absence of SDS. This suggests the possibility of multiple forms of the enzyme. Photoaffinity labelling of bovine liver benzoyltransferase with the active site-directed reagent p-azidobenzoyl-CoA showed that there was only one active site per enzyme molecule (Lau *et al.*, 1977).

In contrast to the bovine liver transferases, a molecular weight of 24,000 has been reported for both benzoyltransferase and phenylacetyltransferase from Rhesus monkey liver, as determined by chromatography on Bio-Gel P-150 (Webster *et al.*, 1976). No data are available concerning the isolation or characterization of the putative taurine N-acyltransferase.

Nandi *et al.* (1979) noted a differential response of the bovine liver benzoyl- and phenylacetyl-transferases to sulphydryl reagents. 5,5'-Dithiobis (2-nitro-benzoic acid) and p-chloromercuribenzoate caused marked inhibition of phenylacetyltransferase activity, whereas pre-incubation with iodoacetamide or

N-ethylmaleimide were without effect. No inhibition of benzoyltransferase activity was found with any of the above reagents, in agreement with previous findings for both the cow (Schachter and Taggart, 1954) and the rat (Brandt *et al.*, 1968) liver enzymes.

Nickel, zinc and, to a lesser extent magnesium ions, all at high non-physiological concentrations, inhibited both transferases in a concentration-dependent manner, which was specific for each divalent cation. Conversely, several monovalent cations, particularly potassium ions, stimulated phenylacetyltransferase activity, which was nearly inactive in the assay buffer alone (25 mM Tris-HCl, pH 8.0). James and Bend (1978a) noted that Tris buffer reacted with phenylacetyl-CoA, and therefore could not be used in their radiochemical Gly.N.A.T. assay; the significance of this interaction, and the mechanism of monovalent cation stimulation have not been established. Benzoyltransferase was not stimulated by monovalent ions; it displayed maximal activity in the assay buffer alone.

From Table 2 it appears that there is a wide variation in the pH optima of Gly.N.A.T. activities from various sources. This may be a reflection of the different assay techniques used and the differences in specific activity of the preparations studied. James and Bend (1978a) reported a value of 9.3 for rabbit kidney Gly.N.A.T., while the enzyme activities from rat (Brandt *et al.*, 1968) and cow (Schachter and Taggart, 1954; Nandi *et al.*, 1979) liver gave values of 8.7 and 8.2–8.6 respectively. Both benzoyl- and phenylacetyltransferase had the same pH optima (8.4–8.6) (Nandi *et al.*, 1979). Ranges of 9–10 noted by Moldave and Meister (1957) for human glutamine N-acyltransferase and 7.7–8.2 for taurine N-acyltransferase from stingray kidney (James, 1978) are the only reports on the effect of pH on these enzymes.

Stability

Gly.N.A.T. from bovine liver mitochondria showed no loss of activity, at various stages of purification, when stored at $-25\,°C$ for five months (Schachter and Taggart, 1954). Both cow enzymes purified by Nandi *et al.* (1979) retained 80% of their activity after one to three weeks storage at 4 °C and benzoyltransferase retained more than half of its activity after several months at $-70\,°C$. Phenylacetyltransferase was, however, inactive following dialysis or concentration by ultrafiltration. A Gly.N.A.T. preparation from rat liver mitochondria (Brandt *et al.* 1968) exhibited appreciable temperature instability. At 38 °C essentially all of the activity was lost in 30 min. At room temperature half of the activity was lost in 24 h. James and Bend (1978a) reported that freezing kidney tissue resulted in a 50% loss in Gly.N.A.T. activity in the rat and rabbit. Caldwell *et al.* (1976), who studied the over-all reaction, reported that hippuric acid formation survived in rat and human cadaver liver for at least 72 h when corpses were stored at 4 °C.

Kinetic properties

Table 2 shows the apparent K_m and V_{max} values obtained for preferred substrates toward various acyl-CoA: amino acid N-acyltransferase preparations. Generally, the apparent K_m values are high for the amino acid acceptors relative to the preferred acyl-CoA substrates ($> 10^{-3}$ M compared to 10^{-5} M range). This suggests that the amino acid availability could determine the extent of conjugation *in vivo*. These kinetic data should, however, be interpreted with caution because of the insensitivity of methods for reliable product detection at acyl-CoA substrate concentrations well below their apparent K_m values. Furthermore, where very high apparent K_m values for both the CoA thioester and amino acid occur, the values shown are only approximate since it is often the case that neither substrate is tested at near saturating concentrations (see Webster *et al.*, 1976).

Recently, Nandi *et al.* (1979) have reported the first preliminary bisubstrate and product inhibition kinetic studies on this group of enzymes. Their results are consistent with a sequential reaction mechanism, wherein the acyl-CoA substrate associates first with enzyme, the amino acid adds to the enzyme prior to dissociation of the first product, namely CoA, and the amino acid conjugate dissociates last. Intersecting lines were obtained on the primary double reciprocal plots which are inconsistent with a double-displacement or ping-pong mechanism (Cleland, 1963). Moreover, a mechanism involving an acyl-enzyme intermediate appears unlikely based on the findings of Lau and coworkers, where photoaffinity-labelled benzoyltransferase contained both the arylacyl and the CoA moieties of p-azidobenzoyl-CoA (Lau *et al.*, 1977). This study additionally provided evidence for the formation of a binary complex between acyl-CoA and the enzyme, in support of the findings that benzoyl-CoA, but not glycine, protected Gly.N.A.T. from heat inactivation (Webster *et al.*, 1974), and that phenylacetyl-CoA, but not glycine, protected phenylacetyltransferase from 5,5′-dithiobis(2-nitrobenzoic acid) inhibition (Nandi *et al.*, 1979). Peptide products should protect these enzymes too.

Marked differences in the reaction rates of benzoyltransferase with different acyl-CoA substrates at near their kinetically saturating concentrations were observed by Nandi *et al.* (1979), arguing against dissociation of CoA from the enzyme as being a final common rate-limiting step. Detailed kinetic studies with non-preferred acyl-CoA substrates, e.g. butyryl-CoA are, however, required to substantiate this.

Effect of age

James and Bend (1978b) have studied the ontogenic development of Gly.N.A.T. Preparations of this enzyme from hepatic mitochondria of rat and rabbit attained adult specific activities by birth and four weeks, respectively, whereas mitochondrial preparations from rabbit kidney did not attain adult levels until four months of age. p-Aminohippurate synthesis was absent or at very low levels in foetal and neonatal hepatic tissue from rats (Brandt, 1966), mice (Gorodischer *et al.*, 1971) and humans (Vest and Salzburg, 1965). In newborn infants, glycine conjugation develops rapidly from birth and reaches adult levels within the first eight weeks of life (Vest and Salzburg, 1965). Alimova (1958—cited in Irjala, 1972) showed the presence of hippurate in the urine of human infants from the second day of life. By contrast, the amount of hippurate excreted in response to a dose of benzoic acid is markedly lower in geriatric subjects as compared to young adults (Stern *et al.*, 1946; Binet *et al.*, 1950, cited in Irjala, 1972). This may be due to reduced amino acid availability in old age, since administration of glycine restores conjugating ability.

Effect of chemical pretreatment

The pretreatment of adult rats or immature rabbits with salicylic acid, benzoic acid or phenobarbitone had little effect on hepatic or renal Gly.N.A.T. activity (James and Bend, 1978b). Irjala (1972), who assayed the over-all conjugation reaction in rat liver and kidney slices following pretreatment of animals with salicylate, found only a 1.5–2-fold increase in the synthesis of salicylglycine from salicylic acid and of 4-aminobenzoylglycine from 4-aminobenzoic acid. Similarly, Parkki (1978) reported that hippurate formation in rat liver is poorly inducible following treatment with a variety of microsomal inducing agents. There is some evidence that the conjugation of organic acids with aspartic acid in plants is inducible (Venis, 1972).

BILE ACID CONJUGATION

In addition to the conjugation of xenobiotics with amino acids, endogenous compounds, i.e. bile acids, conjugate with glycine and taurine. The bile acids are formed from cholesterol, this metabolism occurring only in the liver (Percy-Robb and Boyd, 1973; Danielsson, 1976), and the bile acid conjugates are secreted in bile and undergo enterohepatic circulation.

The reaction mechanism for the conjugation of bile acids is thought to be analogous to that of hippurate synthesis (Elliot, 1956), but exhibits a different subcellular distribution. Choloyl-CoA synthetase (cholate: CoA ligase; EC 6.2.1.7, Elliot, 1957), which catalyses the formation of the activated bile acid intermediate, is a microsomal enzyme in human (Schersten, 1967), rat (Killenberg, 1978) and guinea pig liver (Vessey and Zakim, 1977). A second enzyme, bile acid-CoA: amino acid N-acyltransferase (EC 2.3.1.-) catalyses the final step. The intracellular location of this enzyme(s) has been a matter of debate for some time. Siperstein and Murray (1955) localized the enzyme(s) in the soluble fraction of liver cells, but Bremer (1956) and Elliot (1956) concluded that activity was located in the microsomal fraction and Schersten (1967), in the lysosomal fraction. Recent evidence obtained from bovine (Vessey *et al.*, 1977) and rat liver (Killenberg, 1978) support the original findings of Siperstein and Murray. However, an apparent bimodal distribution of activity, with major fractions of total activity localized in both microsomal and supernatant fractions, is often seen. Killenberg (1978) found that the rat microsomal activity is completely solubilized when microsomes are resuspended in a buffer containing divalent metals, suggesting that the enzyme fractionating with the pellet is adsorbed to the microsomes by coulombic bonds.

Analysis of bile from a large number of animal species has given clear evidence that evolution of bile salts has taken place from C_{27} or C_{26} bile alcohols (conjugated with sulphate), through C_{27} bile acids (conjugated with taurine), to the C_{24} bile acids (conjugated with taurine and/or glycine) found in modern species of fish, reptiles, birds and mammals (Haslewood, 1962, 1964, 1967). Thus taurine conjugation precedes glycine conjugation in evolutionary terms, with all animals except for placental mammals (*Eutheria*), conjugating their bile acids exclusively with taurine. Bile from eutherian mammals can contain either primarily taurine conjugates (dog, rat), primarily glycine conjugates (rabbit), or substantial amounts of both (cow and human). Glycine conjugation is usually found in herbivorous animals, whereas taurine conjugation is predominant in carnivores (cf. the taurine conjugation of arylacetic acids; Idle *et al.*, 1978). Additionally, in humans, the ratio of taurine to glycine conjugates can vary dramatically with the physiological status of the individual (Garbutt *et al.*, 1971), and with age. Thus, at birth, taurine conjugates of bile acids predominate, but the proportion of glycine conjugates increases with age until in adult life the ratio of glycine to taurine conjugates is about three (Murphy and Signer, 1974).

As with conjugation of exogenous acids, little is known of the biochemical basis and significance of the variation in bile acid conjugation between species. Investigators have concentrated on the transferase reaction, where amino acid specificity would be expected to influence the spectrum of conjugates seen. These studies have, until very recently, been limited by the lack of sensitive and specific methods. Prior to 1978, investigations of enzyme specificity included alternative substrate inhibition experiments, which were interpreted as supporting the presence of two (or more) enzymes, one specific for glycine and the other for taurine, (Bremer, 1956; Vessey *et al.*, 1977). More recent reports, however, have described the purification and kinetic characterization in rat (Killenberg and Jordan, 1978) and bovine liver (Vessey, 1979; Czuba and Vessey, 1980) of a single transferase (choloyl-CoA: glycine/taurine N-acyl-transferase), which catalyses the formation of both taurine and glycine conjugates.

Table 3 summarizes of the properties of this soluble enzyme, which is distinct from mitochondrial benzoyltransferase and phenylacetyltransferase. Benzoyl-CoA, phenylacetyl-CoA and acetyl-CoA are not substrates for the bile acid

Table 3 Properties of Choloyl-CoA: glycine/taurine N-acyltransferase

Property	Enzyme source*	
	Bovine liver	Rat liver
1. Subcellular location	soluble fraction	
2. Fold purification	900	200
3. Molecular weight	$\sim 50,000$, probably single polypeptide chain.	—
4. Substrates:		
a. Acyl-CoA	CoA esters of cholic, deoxycholic, chenodeoxycholic and lithocholic acids	
b. Amino acid	glycine, taurine.	glycine, taurine, β-alanine, aminomethanesulphonate.
5. Stability	stable up to 50 °C, stabilized against temperature denaturation by taurocholate.	stable at -70 °C for several weeks, unstable above 48 °C.
6. Inhibitors:		
a. Bile acids and their conjugates	glycocholate, taurocholate, cholate.	—
b. Sulphydryl reagents	p-mercuribenzoate sensitive: reversible and pH dependent; choloyl-CoA and glycine protect. N-ethylmaleimide and iodoacetate insensitive.	DTNB† sensitive, dependent upon concentration and exposure time
7. Optimum pH	—	7.8–8.0
8. Catalytic mechanism	tetra-uni ping-pong; choloyl-CoA thiolase activity at same active site	—

* Bovine data from Vessey (1979); Czuba and Vessey (1980), and rat data from Killenberg and Jordan (1978)
† 5,5′-Dithiobis (2-nitrobenzoic acid)

conjugating enzyme in the presence of glycine or taurine; phenylacetyl-CoA is also inactive with L-glutamine in the rat liver preparation. Moreover, mechanistic studies by Czuba and Vessey (1980) on the soluble bovine enzyme gave strong evidence for a tetra-uni ping-pong mechanism. This contrasts with that for benzoyl- and phenylacetyltransferase activities, which follow a sequential mechanism (see above).

Kinetic data obtained *in vitro* are consistent with the conjugation patterns of bile acids seen *in vivo*. The apparent K_m for the acyl-CoA substrates listed in Table 3 was in the range 10–20 μM for the rat liver enzyme, and was independent of the amino acid acceptor (glycine, 100 mM; taurine, 20 mM). In the same preparation, at 50 μM deoxycholoyl-CoA, the apparent K_m for taurine was

0.8 mM, while the value for glycine was 31 mM. This is consistent with the almost exclusive production of taurine conjugates seen in the rat *in vivo*. Similar findings were reported by Anwer and Hegner (1979) using isolated rat hepatocytes. The cow, by contrast, eliminates significant amounts of both conjugates *in vivo*. Kinetic analysis of the bovine liver enzyme (Czuba and Vessey, 1980) gave apparent K_m values, obtained from secondary double reciprocal plots, of 8.8 mM and 6.7 mM for glycine and taurine, respectively. The apparent K_m for choloyl-CoA determined by analogous methods was 7.2 μM. From these data it appears that the affinity of the bovine enzyme for glycine and taurine are quite similar, corroborating the findings observed *in vivo*.

It is interesting, from the point of view of species differences in conjugation reactions, that the soluble choloyl-CoA: amino acid N-acyltransferase from rat liver had a measurable activity with β-alanine (50 mM), while no conjugation was detectable with the bovine enzyme, even at 125 mM β-alanine. Arginine was not tested as an acyl acceptor in the bovine preparation, and was inactive with purified rat liver enzyme.

The arginine conjugate of cholic acid has been identified in the bile produced by the isolated perfused rat liver (Yousef and Fisher, 1975). Abnormal conjugates of bile acids with ornithine were found in the bile of guinea pigs (Peric-Golia and Jones, 1962) and humans (Peric-Golia and Jones, 1963) infected with *Klebsiella pneumoniae*. Further, following the injection of ^{14}C-labelled ornithine into the guinea pig, some of it was found in the bile conjugated with the bile acids. Studies *in vitro* showed that liver homogenates from both guinea pig and rat conjugated cholic acid with L-ornithine (Peric-Golia and Jones, 1962).

The bile acids are conjugated, therefore, primarily with taurine and/or glycine, but other amino acids, namely arginine and ornithine, can in certain instances undergo conjugation with these endogenous acids. This compares with the amino acid conjugation of xenobiotics, e.g. benzoic acid (see Table 1). The commonest conjugate of this exogenous acid in both vertebrates and invertebrates is benzoylglycine (hippuric acid), but in certain species other amino acid conjugates are found, e.g. the ornithine conjugate (certain birds and reptiles) and the arginine conjugate (certain invertebrates). Both amino groups of ornithine (**20**) form amide bonds with benzoic acid giving dibenzoylornithine, whereas only one amino group of this amino acid appears to be linked to a bile acid in the ornithocholanic acids (Peric-Golia and Jones, 1963).

$$
\begin{array}{c}
\text{COOH} \\
| \\
\text{CHNH}_2 \\
| \\
(\text{CH}_2)_2 \\
| \\
\text{CH}_2\text{NH}_2
\end{array}
$$

(**20**)

Bile acids, namely lithocholic acid (Strange *et al.*, 1977; Hayes *et al.*, 1979) and cholic acid (Hayes *et al.*, 1980), bind to ligandin (Y protein), an hepatic cytosol protein, which binds certain exogenous organic anions, bilirubin, steroids and carcinogens (Litwack *et al.*, 1971) and also possesses glutathione-*S*-transferase activity (see Hirom and Millburn, 1979, 1981). The physiological significance of this bile acid binding is not clear. It appears to be unrelated to intracellular transport; bile acid transit times across the liver are too rapid for them to diffuse across the hepatocyte bound to protein (Strange *et al.*, 1979). Possibly ligandin restricts the partitioning of bile acids into membrane lipid, thereby keeping them in the cytosol, and thus promoting their rapid excretion— as amino acid conjugates produced by the soluble choloyl-CoA : glycine/taurine *N*-acyltransferase—into bile (Hayes *et al.*, 1980). By contrast, the binding of xenobiotic acids to hepatic or renal cytosol proteins, in relation to the effect this could have on their mitochondrial amino acid conjugation and subsequent transport of the conjugate products out of the cell, does not appear to have been investigated.

DIPEPTIDE CONJUGATION

The formation of dipeptide conjugates appears to be a rare metabolic reaction of foreign compounds. In the cat, quinaldylglycyltaurine (21) is a metabolite of quinaldic acid (quinoline-2-carboxylic acid) and of kynurenic acid (4-hydroxy-quinoline-2-carboxylic acid) (Kaihara and Price, 1961). *N*-(3-Phenoxybenzoyl)-glycylvaline (15; see Introduction) is the principal metabolite of 3-phenoxy-benzoic acid, itself a metabolite of the photostable pyrethroid insecticides, in the mallard duck (Huckle *et al.*, 1981a), but has not been found in any other

$$CONHCH_2CONHCH_2CH_2SO_3H$$
(21)

animal species tested so far (Figure 4 and Huckle *et al.*, 1981a). The insecticide DDT forms a dipeptide conjugate containing DDA, aspartic acid and serine in the rat (Pinto *et al.*, 1965). The natural emulsifiers present in the digestive juice of the edible crab, *Cancer pagurus*, are dipeptide conjugates of fatty acids containing sarcosine (*N*-methylglycine) and taurine (cf. bile acid conjugates of vertebrates) (Van den Oord *et al.*, 1965). In the rat, mouse and man, the anti-neoplastic agent methotrexate (4-amino-10-methylpteroylglutamic acid) forms polypeptide conjugates (polyglutamates, cf. folic acid metabolism), as does aminopterin (4-aminopteroylglutamic acid) in the rat (see Israili *et al.*, 1977).

The biochemical mechanism of dipeptide conjugation is unknown. In the formation of the glycylvaline dipeptide conjugate of 3-phenoxybenzoic acid,

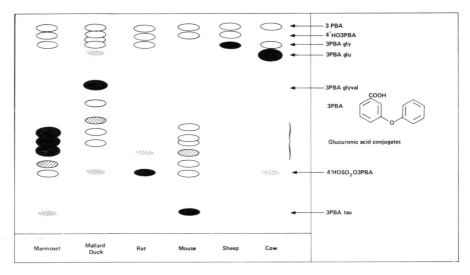

Figure 4   Conjugation of 3-phenoxybenzoic acid (3PBA): metabolites in urine from several species

for example, does the glycine add to the xenobiotic first, followed by valine, or is glycylvaline joined as such to 3-phenoxybenzoic acid? If the latter mechanism occurs, then one might ask: 'What is the normal physiological role of a dipeptide such as glycylvaline in certain avian species?'

CONTROL OF AMINO ACID CONJUGATION

The over-all extent of amino acid conjugation may be affected by a number of factors, such as: enzyme specificity, availability of amino acids and cofactors, membrane permeability to substrates and products. Being a two-step process, the over-all rate will be determined by the slowest step.

Rate-limiting step

For hippurate synthesis in rat liver (Gatley and Sherratt, 1977) and salicylurate synthesis in bovine liver mitochondria (Forman et al., 1971), it appears that formation of the acyl-CoA intermediate is rate-limiting. Similar conclusions were inferred by Vessey (1978) concerning choloyl-CoA synthetase in bile acid conjugation. When each step was determined at nearly optimal substrate concentrations, Forman et al. (1971) found more than a 1000-fold excess of acylating activity as compared to acyl-CoA formation in mitochondrial preparations. Furthermore, the rate of the activating step matched that for the over-all pathway, and the latter was not increased by addition of highly purified transferase enzyme. Recently, both the activating and transferase steps involved in the conjugation of the pyrethroid metabolite 3-phenoxybenzoic acid with glycine have been investigated in hepatic and renal tissues from several mammalian species (Huckle *et al.*, 1981c). In most of the species tested, renal activities of both enzymes were higher than those of corresponding liver preparations. A ten- to 300-fold excess of acyl-CoA: glycine *N*-acyltransferase activity as compared to 3-phenoxybenzoyl-CoA synthetase activity was found, and the rate of the activating step matched that of the over-all conjugation reaction. Therefore, in the formation of 3-phenoxybenzoylglycine, as for hippurate and salicylurate synthesis and bile acid conjugation, the formation of the acyl-CoA thioester intermediate appears to be the rate-limiting step in amino acid conjugation.

How these *in vitro* data compare with the events occurring in the whole animal will depend, however, on comparison of rates of both activating and acylating reactions at physiological concentrations of pathway components. Furthermore, the physiological significance of enzyme activities in disrupted mitochondrial preparations may be uncertain because of ignorance about the conditions in the matrix. It is probable, due to their localization in the mitochondrial matrix, that the two reactions are closely coupled *in vivo*.

Amino acid specificity

Although the activating step appears to regulate the rate of the over-all conjugation reaction, it is at the acylating stage where amino acid specificity would be expected to determine the range of conjugates found *in vivo*. The acyl-CoA: amino acid *N*-acyltransferases are, therefore, particularly important as regards species differences in amino acid conjugation. A variety of amino acids are used by animals for conjugation (e.g. glycine, taurine, glutamine, glutamic acid, ornithine, arginine), and there are probably more inter-species variations in this than in any other detoxication mechanism. The recent isolation of benzoyltransferase and phenylacetyltransferase as separate enzymes offers an explanation at the enzyme level for certain of the species differences in amino acid conjugation observed *in vivo*. Thus, benzoic acid (the principal acyl donor substrate of benzoyltransferase) forms the glycine conjugate (hippuric acid) in primates, whereas phenylacetic acid is conjugated with glutamine in anthropoid species (see, Millburn, 1978). This difference in amino acid conjugation can be explained biochemically by the finding that phenylacetic acid is metabolized by a different transferase to benzoic acid, i.e. by phenylacetyltransferase, which in anthropoid species is specific for glutamine as the amino acid substrate (Webster *et al.*, 1976).

A further example in the metabolism of benzoic and phenylacetic acids concerns their amino acid conjugation in *Chiroptera* (bats). This is the largest order of mammals after the *Rodentia*, containing some 800–900 species of bats.

Table 4   Conjugation in *Chiroptera*

Species	Conjugate formed by: benzoic acid	phenylacetic acid
Indian fruit bat (*Pteropus giganteus*)	1. glucuronide (major) 2. glutamic acid (minor) no benzoylglycine	1. glycine (major) 2. glucuronide (minor)
African fruit bat (*Epomops franqueti*)	1. glycine 2. glucuronide 3. glutamic acid (minor)	—
British bat (*Pipistrellus pipistrellus*)	glycine	glycine
Vampire bat (*Desmodus rotundus*)	—	glycine

Data taken from Millburn (1978) and references therein

The conjugation of benzoic acid and phenylacetic acid in several species belonging to the Order *Chiroptera* is shown in Table 4. The Indian fruit bat is the only species of mammal known to be defective in the glycine conjugation of benzoic acid, which in this bat is metabolized to benzoylglucuronide, although it can form phenylacetylglycine. Indian and African fruit bats can also form benzoylglutamic acid as a minor metabolite of benzoic acid. Since the Indian fruit bat can conjugate phenylacetic acid with glycine and benzoic acid with glutamic acid, its inability to form benzoylglycine would appear to result from a defect in benzoyltransferase rather than from a lack of benzoyl-CoA formation. Phenylacetyltransferase is functioning normally in this species of bat resulting in the production of phenylacetylglycine as the major metabolite of phenylacetic acid.

The species difference in the metabolism of benzoic acid by the Indian fruit bat appears, therefore, to result from the lack of benzoyltransferase activity. This could be due to the presence of an inactive enzyme protein or lack of synthesis of the enzyme. Although lack of enzyme activity results in striking species differences in metabolism, the commonest causes of such differences are probably variations in the rate of enzymic action. Thus, in the conjugation of the pyrethroid metabolite 3-phenoxybenzoic acid, the gerbil and ferret, which excrete 3-phenoxybenzoylglycine as the principal urinary metabolite, have higher levels of 3-phenoxybenzoyl-CoA synthetase and glycine N-acyltransferase activity (measured *in vitro*) than the hamster, which excretes only small amounts of the glycine conjugate (Huckle *et al.*, 1981c). By contrast, the mouse, which eliminates 3-phenoxybenzoic acid in the urine as the taurine conjugate (Hutson and Casida, 1978), has good enzymic capacity for the conjugation of this xenobiotic with glycine *in vitro* (Huckle *et al.*, 1981c). Conjugation of 3-phenoxybenzoic acid with taurine has not yet been demonstrated *in vitro*, but two possibilities can be considered. Firstly, are there two distinct acyl-CoA: amino acid N-acyltransferases, one specific for glycine and the other for taurine, the latter enzyme being the more active of the two? Secondly, mouse tissues may contain a single transferase, which can utilize both glycine and taurine as amino acid substrates, the K_m for glycine being higher than that for taurine. 3-Phenoxybenzoic acid can also be conjugated with glutamic acid (in the cow; see Figure 4); cf. the formation of benzoylglutamic acid in bats (Table 4). Nothing is known concerning the biochemical mechanism for the conjugation of xenobiotics with glutamic acid. Is there a separate transferase for this amino acid?

The extent to which a xenobiotic substrate is conjugated with an amino acid *in vivo* may also be influenced by the extent to which other competing metabolic pathways operate. For example, the rat excretes 3-phenoxybenzoic acid mainly as a sulphate conjugate in urine (Figure 4), but the liver and kidneys of this species have the enzymic capacity to conjugate 3-phenoxybenzoic acid with glycine *in vitro* (Huckle *et al.*, 1981c.) Possibly the extensive hydroxylation of 3-phenoxybenzoic acid by the rat *in vivo* (Crayford and Hutson, 1980; Huckle *et al.*, 1981a), exceeds the capacity of the tissues to conjugate this acid with glycine. This would account for the small amounts of 3-phenoxybenzoylglycine produced by the rat *in vivo*. The hydrolysis of amino acid conjugates in tissues may also be a factor in the over-all metabolism of a xenobiotic. In a recent study on the metabolism of salicylic acid in the isolated perfused rat kidney (Bekersky *et al.*, 1980a), its glycine conjugate, salicyluric acid, was found to be hydrolysed by the renal tissue. Furthermore, the salicylic acid formed by this hydrolysis was cleared by the kidneys at a different rate than administered salicylic acid, presumably due to rate-limiting diffusion of salicylic acid into the renal cells (Bekersky *et al.*, 1980b). Rat liver does not possess this hydrolytic activity towards salicyluric acid (Bekersky *et al.*, 1980a).

Membrane permeability to substrates and products

Consideration of substrate and product compartmentation and transport into and out of mitochondria may be important in the amino acid conjugation of xenobiotics. Glycine, ammonium benzoate or ammonium hippurate (150 mM solutions) appear to travel freely across the inner membrane of mitochondria from rat liver (McGivan and Klingenberg, 1971). Halling *et al.* (1975) concluded that glycine crosses the inner membrane as the zwitterion, while it is envisaged that undissociated benzoic acid and hippuric acid pass through due to their lipid solubility. There is no evidence, at least with components concerned with hippurate synthesis, of the involvement of specific carriers in the inner mitochondrial membrane in rat liver. Thus, benzoyl-CoA does not appear to be a substrate for carnitine acyltransferase, which catalyses the conjugation of the carrier molecule carnitine with natural long-chain saturated fatty acids (Gatley and Sherratt, 1977). Acyl carnitine conjugates are then transported across the inner mitochondrial membrane to the site of fatty acid β-oxidation. Carrier mediated transport systems for some neutral amino acids (Le Cam and Freychet, 1977; Kelley and Potter, 1978), and taurine (Hardison and Weiner, 1979) have, however, been reported in isolated hepatocytes and it is conceivable that regulation of conjugation in these cells may be partly exerted through such systems. Nevertheless, although the transport of amino acids across biological membranes has been investigated for a number of years, the transport mechanism remains unknown. The movement of most amino acids across mammalian cell membranes appears to be an active process involving a mobile carrier molecule(s), which binds the amino acid prior to translocation (see McNamara and Ozegovic, 1980). Evidence with bile acids (Anwer and Hegner, 1979) would suggest that in hepatocytes the uptake rate of cholic acid is not dependent on, and thus not regulated by, its conjugation.

Gatley and Sherratt (1977) have summarized the events believed to occur in the synthesis of hippurate in rat liver mitochondria. Benzoic acid crosses the inner mitochondrial membrane into the matrix and reacts with CoA forming benzoyl-CoA. This reaction results in the hydrolysis of ATP to AMP and pyrophosphate (PP_i). The latter hydrolysis product is converted by inorganic pyrophosphatase (EC 3.6.1.1) into phosphate (P_i). The AMP is rephosphorylated to ADP by GTP, which is converted to GDP. This is then rephosphorylated by ATP. Thus, the biosynthesis of every molecule of benzoyl-CoA requires the hydrolysis of two molecules of ATP to ADP. Benzoyl-CoA reacts with glycine, which freely crosses the inner mitochondrial membrane (see above), producing hippurate. Although there appears to be no barrier to the movement of hippurate out of mitochondria, with more polar substrates and products (for example, 3-phenoxybenzoyltaurine) poor lipid solubility may significantly affect their passive mobility across mitochondrial membranes.

Availability of amino acids and cofactors

With various acyl-CoA: amino acid N-acyltransferases, the finding of high apparent K_m values for preferred amino acid substrates relative to the preferred acyl-CoA substrates ($> 10^{-3}$ M compared to 10^{-5} M range; see Table 2), suggests that the availability of amino acids could determine the extent of conjugation *in vivo*. The descriptive kinetic data obtained by Webster *et al.* (1976) indicated that the rate of conjugate formation under physiological conditions was probably first order with respect to acceptor amino acid concentration. Thus, the concentration of amino acid in the matrix of the mitochondrion must be high for maximum rates of conjugation.

These data are supported by the findings of Amsel and Levy (1969), who concluded that the concentration of glycine available *in vivo* limits the rate of conjugation and hence elimination of carboxylic acids such as benzoic acid. Further, conjugation of benzoate and phenylacetate with glycine increases *in vivo* with exogenous doses of either acid, especially if glycine is administered concomitantly (Wan and Riegelman, 1972; Amsel and Levy, 1970). At very high doses of benzoate, if glycine is not also administered, excretion of hippurate reaches a plateau; that is, the rate of glycine conjugation is limited. This is not the case, however, with salicylic acid (Levy, 1971).

Taurine conjugation of xenobiotics (with phenylacetyl-CoA as substrate) appears to be located, like glycine conjugation, in the mitochondrial matrix of rat liver and kidneys (James, 1978). The renal enzyme activities were higher than those of hepatic tissue. The sources and availability of taurine *in vivo* could, therefore, be factors in conjugations involving this amino acid.

Taurine has a wide distribution in animals (see Jacobsen and Smith, 1968) and its biosynthesis in mammals proceeds by two routes (Figure 5); the cysteamine pathway and the cysteine sulphinic acid pathway (see Scandurra *et al.*, 1978). The cysteamine pathway involves the indirect conversion of cysteine to cysteamine and starts with pantothenic acid (**22** in Figure 5), which is formed by plants and bacteria but is required in the diet by mammals. Pantothenic acid is phosphorylated by pantothenate kinase (enzyme i in Figure 5, EC 2.7.1.33) to 4'-phosphopantothenic acid (**23**). Cysteine then adds to (**23**), this reaction being catalysed by 4'-phosphopantothenoyl-L-cysteine synthetase (ii, EC 6.3.2.5). The product, 4'-phosphopantothenoyl-L-cysteine (**24**), is converted by the decarboxylase (iii, EC 4.1.1.36) to 4'-phosphopantothenoyl-cysteamine (4'-phosphopantetheine, (**25**)), which is dephosphorylated by alkaline phosphatase (iv, EC 3.1.3.1) to pantothenoylcysteamine (pantetheine, **26**). The enzyme pantetheinase (v, EC 3.5.1.-) splits (**26**) into (**22**) and cysteamine (**27**). This amine cannot be produced directly from cysteine in mammals because of a lack of cysteine decarboxylase activity. Cysteamine is then oxidized to hypotaurine (**28**) by cysteamine dioxygenase (vi, EC 1.13.11.19). Hypotaurine is oxidized to taurine (**29**) by an unidentified enzyme (vii).

The cysteine sulphinic acid pathway begins with the oxidation of cysteine (**30**) to cysteine sulphinic acid (**31**) by cysteine dioxygenase (viii, EC 1.13.11.20). In the rat, this enzyme is located only in the liver. Cysteine sulphinic acid is metabolized to hypotaurine by a specific decarboxylase (ix, EC 4.1.1.29). More $^{14}CO_2$ is produced *in vitro* from ^{14}C-labelled cysteine bound as ^{14}C-labelled phosphopantothenoylcysteine than from ^{14}C-labelled cysteine itself, suggesting that the cysteamine pathway is quantitatively more important than the cysteine sulphinic acid pathway for taurine biosynthesis (Scandurra *et al.*, 1977). However, enzymes (i), (ii) and (iii) in Figure 5 are absent from rat-liver mitochondria (Skrede and Halvorsen, 1979), but are present in the cytosol (see Abiko, 1975). Therefore, since the decarboxylase (iii) is only found in the cytosol, this suggests that mitochondrial 4'-phosphopantothenoyl-L-cysteine (**24**) is not a source of taurine for amino acid conjugation. If taurine is synthesized by the cysteamine pathway in the cytosol, is there a carrier mechanism for the transport of this very polar substrate into the mitochondria to the site of amino acid conjugation?

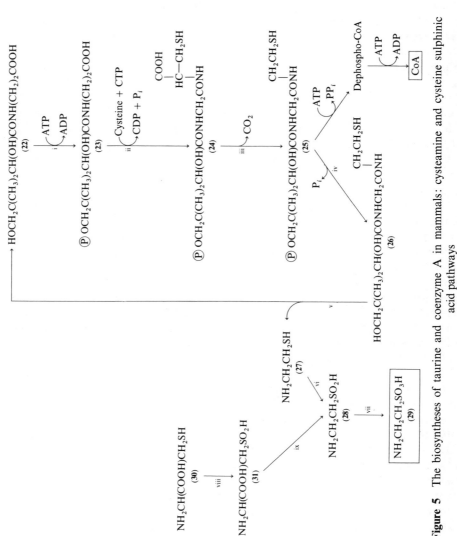

Figure 5 The biosyntheses of taurine and coenzyme A in mammals: cysteamine and cysteine sulphinic acid pathways

Other alternatives to cysteine as a source of cysteamine and, therefore, taurine are possibly (see Figure 6 and Scandurra *et al.*, 1978): (a) cysteine sulphonate (32), which on decarboxylation gives cysteamine sulphonate (33); this, in the presence of a thiol, forms cysteamine; (b) the S-containing amino acid L-lanthionine (34), which is found in animals and can be converted by way of aminoethylcysteine (35) to cysteamine; (c) S-adenosylcysteamine, which is a metabolite of S-adenosylcysteine; and (d) aminoacrylate (36), derived from serine dehydration, which together with sulphate from 3'-phosphoadenosine-5'-phosphosulphate (PAPS) forms cysteic acid (37). Decarboxylation of this intermediate gives taurine. There are, therefore, several sources of taurine in animals, and the extent to which these metabolic pathways function *in vivo* may be of significance in determining the availability of this amino acid for conjugation reactions.

If the supply of amino acids or the rate of their conjugation became limiting, this may help to explain the existence of mitochondrial acyl-CoA hydrolase

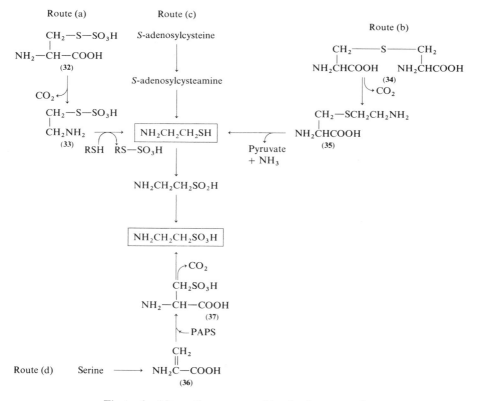

Figure 6   Alternative sources of taurine in mammals

activities of high K_m (Schachter and Taggart, 1954; Gatley and Sherratt, 1976; James and Bend, 1978a). The presence of a 'futile cycle', i.e. simultaneous existence of acyl-CoA synthetase and acyl-CoA hydrolase (EC 3.1.2.1 and 3.1.2.2) activities, may be to prevent the accumulation of acyl-CoA sequestering too much of the available CoA, and thus impairing other CoA-requiring reactions (see Osmundsen and Sherratt, 1975). The availability of mitochondrial CoA could, therefore, be a factor in the conjugation of xenobiotics with amino acids.

The metabolic pathway for the biosynthesis of CoA from pantothenic acid is shown in Figure 5. The first three steps in the pathway, from pantothenic acid (22) to 4'-phosphopantothenoylcysteamine (25), are common to both CoA and taurine biosynthesis (see above). In CoA formation, the intermediate (25) reacts with ATP giving dephospho-CoA. A further ATP molecule is then used to phosphorylate the 3'-OH group of the adenosine part of dephospho-CoA giving the nucleotide derivative, coenzyme A. All the enzymes required for the biosynthesis of CoA from pantothenic acid are present in the cytosol fraction of rat liver (see Abiko, 1975). The inner mitochondrial membrane is not permeable to CoA (Skrede and Bremer, 1970), however, rat-liver mitochondria are able to synthesize CoA from 4'-phosphopantothenoylcysteamine (25) but not from intermediates (23 and 24) or pantothenic acid (22) itself (Skrede and Halvorsen, 1979).

CONCLUDING REMARKS

Amino acid conjugation is a two step, energy (ATP)-dependent biosynthetic reaction, in which the xenobiotic is first activated with CoA to form a nucleotide intermediate. This acyl-CoA thioester then reacts with one of a variety of amino acids to form a conjugate. Two families of enzymes are involved in the over-all reaction: the acyl-CoA synthetases and the acyl-CoA: amino acid N-acyltransferases. With the limited number of substrates examined so far, the activating step appears to be the rate-limiting step in amino acid conjugation. The transferases specify the amino acid utilized for conjugation, and two of these enzymes have been isolated and purified to near homogeneity. Benzoyltransferase is specific for benzoyl-CoA, salicyl-CoA and certain fatty acyl-CoA esters as acyl donors with glycine as the amino acid substrate. Phenylacetyltransferase on the other hand is specific for phenylacetyl-CoA and related arylacetyl-CoA esters with glycine (or glutamine, depending on species) as the acyl acceptor. The isolation and characterization of further enzymes from this transferase family will bring about a greater understanding of the biochemical mechanisms involved in amino acid conjugations, and possibly offer biochemical explanations for species differences in this conjugation reaction. In particular, there is at present little knowledge on the biochemistry of amino acid conjugations other than those involving glycine.

Finally, although this chapter has considered the amino acid conjugations as one of the classical detoxication mechanisms, another biological role for amino acid conjugates has been suggested recently. Hippuric acid may function in the modulation of γ-glutamyl transpeptidase activity *in vivo* (Thompson and Meister, 1980). This enzyme, a glycoprotein, is abundant in the brush border membrane of kidney tubules and small intestine. It catalyses the transfer of the γ-glutamyl moiety of glutathione to various amino acid acceptors, and appears to be involved, via the γ-glutamyl cycle, in the transport of amino acids across the intestinal mucosa and their reabsorption in the kidney tubules (see Meister and Tate, 1976). γ-Glutamyl transpeptidase also has a role in detoxication; its action results in the formation of the cysteinylglycine intermediate from the glutathione conjugate during mercapturic acid biosynthesis. Hippuric acid may regulate γ-glutamyl transpeptidase activity by interacting with the cysteinyl-glycine binding site of the enzyme. Furthermore, although hippuric acid is usually thought of as a detoxication product of benzoic acid, Thompson and Meister (1980) have shown, by using germ-free rats fed a benzoate-free diet, that hippuric acid can be formed endogenously from phenylalanine (via phenylpyruvate, phenyllactate and cinnamate followed by β-oxidation to benzoate). They suggest that this endogenously formed hippurate regulates γ-glutamyl transpeptidase *in vivo* i.e. an alternative physiological function to detoxication for the glycine conjugation of benzoic acid. It is conceivable that other glycine conjugates, usually considered to be detoxication products, also bind to γ-glutamyl transpeptidase initiating interactions of biological significance.

REFERENCES

Aas, M., and Bremer, J. (1968) 'Short-chain fatty acid activation in rat liver. A new assay procedure for the enzymes and studies on their intracellular localisation', *Biochim. Biophys. Acta*, **164**, 157–166.

Abiko, Y. (1975) 'Metabolism of coenzyme A', in *Metabolic Pathways*, (D. M. Greenberg, Ed.), 3rd edn. Academic Press, New York, Vol. 7, pp. 1–25.

Alimova, M. M. (1958) 'Hippuric acid content of the urine of infants during the first year', *Vop. Med. Khim.*, **4**, 280–284.

Amsel, L. P., and Levy, G. (1969) 'Drug biotransformation interactions in man. II. A pharmacokinetic study of the simultaneous conjugation of benzoic and salicyclic acids with glycine' *J. Pharm. Sci.*, **58**, 321–325.

Amsel, L. P., and Levy, G. (1970) 'Effect of ethanol on conjugation of benzoate and salicylate with glycine in man', *Proc. Soc. Exp. Biol. Med.*, **135**, 813–816.

Andreae, W. A., and Good, N. E. (1961) 'Studies on 3-indoleacetic acid metabolism. IV. Conjugation with aspartic acid and ammonia as processes in the metabolism of carboxylic acids', *Plant Physiology*, **32**, 566–572.

Anwer, M. S., and Hegner, D. (1979) 'Study of cholic acid conjugation in isolated rat hepatocytes', *Hoppe-Seyler's Z. Physiol. Chem.*, **360**, 515–522.

Arjmand, M., Hamilton, R. H., and Mumma, R. O. (1978) 'Metabolism of 2,4,5-tri-chlorophenoxyacetic acid. Evidence for amino acid conjugates in soybean callus tissue', *J. Agr. Fd. Chem.*, **26**, 1125–1128.

Bar-Tana, J., and Rose, G. (1968a) 'Studies on medium-chain fatty acyl-coenzyme A synthetase. Enzyme fraction I: Mechanism of reaction and allosteric properties', *Biochem. J.*, **109**, 275–282.

Bar-Tana, J., and Rose, G. (1968b) 'Studies on medium-chain fatty acyl-coenzyme A synthetase. Enzyme fraction. II: Mechanism of reaction and specific properties', *Biochem. J.*, **109**, 283–292.

Bar-Tana, J., Rose, G., and Shapiro, B. (1968) 'Studies on medium-chain fatty acyl-coenzyme A synthetase. Purification and properties', *Biochem. J.*, **109**, 269–274.

Bartlett, K., and Gompertz, D. (1974) 'The specificity of glycine-N-acylase and acylglycine excretion in the organic-acidaemias', *Biochemical Medicine*, **10**, 15–23.

Bekersky, I., Colburn, W. A., Fishman, L., and Kaplan, S. A. (1980a) 'Metabolism of salicylic acid in the isolated perfused rat kidney. Interconversion of salicyluric and salicylic acids', *Drug Metab. Dispos.*, **8**, 319–324.

Bekersky, I., Fishman, L., Kaplan, S. A., and Colburn, W. A. (1980b) 'Renal clearance of salicylic acid and salicyluric acid in the rat and in the isolated perfused rat kidney', *J. Pharmacol. Exp. Ther.*, **212**, 309–314.

Berg, P. (1956) 'Acyladenylates: an enzymatic mechanism of acetate activation', *J. Biol. Chem.*, **222**, 991–1013.

Binet, L., Bourliere, F., and Coulland, D. (1950) 'La diminution avec l'âge de l'hippuricurie provoquée chez l'homme', *C.R. Acad. Sci. (Paris)*, **230**, 698–700.

Borsook, H., and Dubnoff, J. W. (1947) 'Synthesis of hippuric acid in liver homogenate', *J. Biol. Chem.*, **168**, 397–398.

Brandt, I. K. (1966) 'Glycine acyltransferase activity in developing rat liver', *Biochem. Pharmacol.*, **15**, 994–995.

Brandt, I. K., Simmons, P., and Gutfinger, T. (1968) 'Rat glycine acyltransferase: partial purification and some properties', *Biochim. Biophys. Acta*, **167**, 196–198.

Bremer, J. (1956) 'Species differences in the conjugation of free bile acids with taurine and glycine', *Biochem. J.*, **63**, 507–513.

Bridges, J. W., French, M. R., Smith, R. L., and Williams, R. T. (1970) 'The fate of benzoic acid in various species', *Biochem. J.*, **118**, 47–51.

Bridges, J. W., Evans, M. E., Idle, J. R., Millburn, P., Osiyemi, F. O., Smith, R. L., and Williams, R. T. (1974) 'The conjugation of indolylacetic acid in man, monkeys and other species', *Xenobiotica*, **4**, 645–652.

Caldwell, J. (1980) 'Comparative aspects of detoxication in mammals', in *Enzymatic Basis of Detoxication*, (W. B. Jakoby, Ed.), Academic Press, New York, Vol. 1, pp. 85–114.

Caldwell, J., Moffatt, J., and Smith, R. L. (1976) 'Post-mortem survival of hippuric acid formation in rat and human cadaver tissue samples', *Xenobiotica*, **6**, 275–280.

Carter, M. C. (1965) 'Studies on the metabolic activity of 3-amino-1,2,4-triazole', *Physiol. Plant*, **18**, 1054–1058.

Chantrenne, H. (1951) 'The requirement for coenzyme A in the enzymatic synthesis of hippuric acid', *J. Biol. Chem.*, **189**, 227–233.

Cleland, W. W. (1963) 'The kinetics of enzyme-catalysed reactions with two or more substrates or products. I. Nomenclature and rate equations', *Biochim. Biophys. Acta*, **67**, 104–137.

Climie, I. J. G., and Hutson, D. H. (1979) 'Conjugation reactions with amino acids including glutathione', in *Advances in Pesticide Science*, (H. Geissbühler, Ed.), Pergamon Press, Oxford, part 3, pp. 537–546.

Cohen, P. P., and McGilvery, R. W. (1946) 'Peptide bond synthesis. I. The formation of p-aminohippuric acid by rat liver slices', *J. Biol. Chem.*, **166**, 261–272.

Cohen, P. P., and McGilvery, R. W. (1947a) 'Peptide bond synthesis. II. The formation of p-aminohippuric acid by liver homogenates', *J. Biol. Chem.*, **169**, 119–136.

Cohen, P. P., and McGilvery, R. W. (1947b) 'Peptide bond synthesis. III. On the mechanism of p-aminohippuric acid synthesis', *J. Biol. Chem.*, **171**, 121–133.

Cohen, P. P., and Ochoa, S. (1951) 'Utilization of phosphate bond energy in biological systems. The synthesis of peptide bonds' in *Phosphorus metabolism*, (W. D. McElroy and B. Glass, Eds.), John Hopkins Press, Baltimore, Vol. 1, pp. 630–640.

Conti, A., and Bickel, M. H. (1977) 'History of drug metabolism: discoveries of the major pathways in the 19th century', *Drug Metab. Rev.*, **6**, 1–50.

Crayford, J. V., and Hutson, D. H. (1980) 'The metabolism of 3-phenoxybenzoic acid and its glucoside conjugate in rats', *Xenobiotica*, **10**, 355–364.

Czuba, B., and Vessey, D. A. (1980) 'Kinetic characterization of choloyl-CoA', *J. Biol. Chem.*, **255**, 5296–5299.

Danielsson, H. (1976) 'Bile acid metabolism and its control', in *The Hepatobiliary System* (W. Taylor, Ed.), Plenum, New York, pp. 389–404.

De Duve, C., Pressman, B. C. Gianetto, R., Wattiaux, R., and Applemans, F. (1955) 'Tissue fractionation studies (6). Intracellular distribution patterns of enzymes in rat liver tissue', *Biochem. J.*, **60**, 604–617.

Elliot, W. H. (1956) 'The enzymic synthesis of taurocholic acid: a qualitative study', *Biochem. J.*, **62**, 433–436.

Elliot, W. H. (1957) 'The breakdown of adenosine triphosphate accompanying cholic acid activation by guinea pig liver microsomes', *Biochem. J.*, **65**, 315–321.

Esaac, E. G., and Casida, J. E. (1968) 'Piperonylic acid conjugates with alanine, glutamate, glutamine, glycine, serine in living houseflies', *J. Insect. Physiol.*, **14**, 913–925.

Farrar, W. W., and Plowman, K. M. (1979) 'Kinetics of acetyl-CoA synthetase. II. Product inhibition studies', *Int. J. Biochem.*, **10**, 583–588.

Feung, C. S., Hamilton, R. H., and Mumma, R. O. (1973) 'Metabolism of 2,4-dichloro-phenoxyacetic acid. V. Identification of metabolites in soyabean callus tissue cultures', *J. Agr. Fd. Chem.*, **21**, 637–640.

Forman, W. B., Davidson, E. D., and Webster, L. T., Jr. (1971) 'Enzymatic conversion of salicylate to salicylurate', *Molecular Pharmacology*, **7**, 247–259.

Garbutt, J. T., Lack, L., and Tyor, M. P. (1971) 'Physiological basis of alterations in the relative conjugation of bile acids with glycine and taurine', *Am. J. Clin. Nutr.*, **24**, 218–228.

Garland, P. B., Yates, D. W., and Haddock, B. A. (1970) 'Spectrophotometric studies of acyl-coenzyme A synthetases of rat liver mitochondria', *Biochem. J.*, **119**, 553–564.

Gatley, S. J. (1979) 'A specific enzymatic assay for glycine', *Anal. Lett.*, **12**, 415–419.

Gatley, S. J., and Sherratt, H. S. A. (1976) 'The localization of hippurate synthesis in the matrix of rat liver mitochondria', *Biochem. Soc. Trans.*, **4**, 525–526.

Gatley, S. J., and Sherratt, H. S. A. (1977) 'The synthesis of hippurate from benzoate and glycine by rat liver mitochondria. Submitochondrial localisation and kinetics', *Biochem. J.*, **166**, 39–47.

Gaughan, L. C., Unai, T., and Casida, J. E. (1977) 'Permethrin metabolism in rats', *J. Agr. Fd. Chem.*, **25**, 9–17.

Gaughan, L. C., Ackerman, M. E., Unai, T., and Casida, J. E. (1978a) 'Distribution and metabolism of *trans*- and *cis*-permethrin in lactating Jersey cows', *J. Agr. Fd. Chem.*, **26**, 613–618.

Gaughan, L. C., Robinson, R. A., and Casida, J. E. (1978b) 'Distribution and metabolic fate of *trans*- and *cis*-permethrin in laying hens', *J. Agr. Fd. Chem.*, **26**, 1374–1380.

Gingell, R. (1976) 'Metabolism of ^{14}C-DDT in the mouse and hamster', *Xenobiotica*, **6**, 15–20.

Gianfanga, T. J., and Davies, P. J. (1980) '*N*-Benzoylaspartate and *N*-phenylacetylaspartate from pea seeds', *Phytochem.*, **19**, 959–961.

Glickman, A. H., Shono, T., Casida, J. E., and Lech, J. J. (1979) '*In vitro* metabolism of permethrin isomers by carp and rainbow trout liver microsomes', *J. Agr. Fd. Chem.*, **27**, 1038–1041.

Gorodischer, R., Kranser, J., and Yaffe, S. J. (1971) 'Hippuric acid synthesising system during development. Kinetic studies and inhibition with salicylic acid', *Biochem. Pharmacol.*, **20**, 67–72.

Graham, A. B., and Park, M. V. (1969) 'A product-inhibition study of the mechanism of mitochondrial octanoyl-coenzyme A synthetase', *Biochem. J.*, **111**, 257–262.

Groot, P. H. E., and Scheek, L. M. (1976) 'Acyl-CoA synthetases in guinea pig liver mito-chondria. Purification and characterisation of a distinct propionyl-CoA synthetase', *Biochim. Biophys. Acta*, **441**, 260–267.

Groot, P. H. E., Van Loon, C. M. I., and Hülsmann, W. C. (1974) 'Identification of the palmitoyl-CoA synthetase present in the inner membrane-matrix fraction of rat liver mitochondria', *Biochim. Biophys. Acta*, **337**, 1–12.

Groot, P. H. E., Scholte, H. R., and Hülsmann, W. C. (1976) 'Fatty acid activation: Specificity, localization and function', *Adv. Lipid Res.*, **14**, 75–126.

Grunow, W., and Böhme, Chr. (1974) 'Metabolism of 2,4,5-T and 2,4-D in rats and mice', *Arch. Toxicol.*, **32**, 217–225.

Guarino, A. M., James, M. O., and Bend, J. R. (1977) 'Fate and distribution of the herbi-cides 2,4-D and 2,4,5-T in the dogfish shark', *Xenobiotica*, **7**, 623–631.

Hahlbrock, K., and Grisebach, H. (1970) 'Formation of coenzyme A esters of cinnamic acids with an enzyme preparation from cell suspension cultures of parsley', *FEBS Lett.*, **11**, 62–64.

Halling, P. J., Brand, M. D., and Chappell, J. B. (1975) 'Permeability of mitochondria to neutral amino acids', *FEBS Lett.*, **34**, 169–171.

Hardison, W. G., and Weiner, R. G. (1979) 'Characteristics of taurine transport in primary cultures of rat hepatocytes', *Gastroenterology*, **76**, 1283.

Haslewood, G. A. D. (1962) 'Bile salts: structure, distribution and possible biological significance as a species character', in *Comparative Biochemistry*, (M. Florkin and H. S. Mason, Eds.), Academic Press, New York, Vol. 3, pp. 205–229.

Haslewood, G. A. D. (1964) 'The biological significance of chemical differences in bile salts', *Biol. Rev.*, **39**, 537–574.

Haslewood, G. A. D. (1967) *Bile Salts*, Methuen, London.

Hayes, J. D., Strange, R. C., and Percy-Robb, I. W. (1979) 'Identification of two litho-cholic acid-binding proteins. Separation of ligandin from glutathione S-transferase B', *Biochem. J.*, **181**, 699–708.

Hayes, J. D., Strange, R. C., and Percy-Robb, I. W. (1980) 'Cholic acid binding by gluta-thione S-transferase from rat liver cytosol', *Biochem. J.*, **185**, 83–87.

Hirom, P. C., Idle, J. R., and Millburn, P. (1977) 'Comparative aspects of the biosyn-thesis and excretion of xenobiotic conjugates by non-primate mammals', in *Drug Metabolism—from Microbe to Man*, (D. V. Parke and R. L. Smith, Eds.), Taylor and Francis, London, pp. 299–329.

Hirom, P. C., and Millburn, P. (1979) 'Enzymic mechanisms of conjugation', in *Foreign Compound Metabolism in Mammals*, (D. E. Hathway, Ed.), The Chemical Society, London, Vol. 5, pp. 132–158.

Hirom, P. C., and Millburn, P. (1981) 'Enzymic mechanisms of conjugation' in *Foreign Compound Metabolism in Mammals*, (D. E. Hathway, Ed.), The Chemical Society, London, Vol. 6, pp. 111–132.

Huckle, K. R., Climie, I. J. G., Hutson, D. H., and Millburn, P. (1981a) 'Dipeptide conjugation of 3-phenoxybenzoic acid in the mallard duck', *Drug Metab. Dispos.*, **9**, 147–149.

Huckle, K. R., Hutson, D. H., and Millburn, P. (1981b) 'Species differences in the metabolism of 3-phenoxybenzoic acid', *Drug Metab. Dispos.*, **9**, 352–359.

Huckle, K. R., Tait, G. H., Millburn, P., and Hutson, D. H. (1981c) 'Species variations in the renal and hepatic conjugation of 3-phenoxybenzoic acid with glycine', *Xenobiotica*, **11**, 635–644.

Hutson, D. H. (1979) 'The metabolic fate of synthetic pyrethroid insecticides in mammals', in *Progress in Drug Metabolism* (J. W. Bridges, and L. F. Chaussaud, Eds.), John Wiley, Chichester, Vol. 3, pp. 215–252.

Hutson, D. H. (1981) 'The metabolism of insecticides in man', in *Progress in Pesticide Biochemistry*, (D. H. Hutson and T. R. Roberts, Eds.), John Wiley, Chichester, Vol. 1, pp. 287–333.

Hutson, D. H., and Casida, J. E. (1978) 'Taurine conjugation in metabolism of 3-phenoxybenzoic acid and the pyrethroid insecticide cypermethrin in mouse', *Xenobiotica*, **8**, 565–571.

Hutson, D. H., and Pickering, B. A. (1971) 'The metabolism of isopropyl oxitol in rat and dog', *Xenobiotica*, **1**, 105–119.

Idle, J. R., Millburn, P., and Williams, R. T. (1978) 'Taurine conjugates as metabolites of arylacetic acids in the ferret', *Xenobiotica*, **8**, 253–264.

Irjala, K. (1972) 'Synthesis of *p*-aminohippuric, hippuric and salicyclic acids in experimental animals and man', *Ann. Acad. Sci. Fenn. Ser. A. V.*, **154**, 1–40.

Israili, Z. H., Dayton, P. G., and Kiechel, J. R. (1977) 'Novel routes of drug metabolism: A survey', *Drug Metab. Dispos.*, **5**, 411–415.

Ivie, G. W., and Hunt, L. M. (1980) 'Metabolites of *cis*- and *trans*-permethrin in lactating goats', *J. Agr. Fd. Chem.*, **28**, 1131–1138.

Jacobsen, J. G., and Smith, L. H. (1968) 'Biochemistry and physiology of taurine and taurine derivatives', *Physiol Rev.*, **48**, 424–511.

James, M. O. (1978) 'Taurine conjugation of carboxylic acids in some marine species' in *Conjugation reactions in drug biotransformation* (A. Aitio, Ed.), Elsevier, North-Holland Biomedical Press, Amsterdam, pp. 121–129.

James, M. O., and Bend, J. R. (1978a) 'A radiochemical assay for glycine *N*-acyltransferase activity', *Biochem. J.*, **172**, 285–291.

James, M. O., and Bend, J. R. (1978b) 'Perinatal development of, and effects of chemical pretreatment on, glycine *N*-acyltransferase activities in liver and kidney of rabbit and rat', *Biochem. J.*, **172**, 293–299.

James, M. O., Smith, R. L., Williams, R. T., and Reidenberg, M. (1972) 'The conjugation of phenylacetic acid in man, sub-human primates and some non-primate species', *Proc. R. Soc. London Ser. B.*, **182**, 25–35.

Jenks, W. P. (1962) 'Acyl activation' in *The Enzymes*, (P. D. Boyer, Ed.), 2nd edn., Academic Press, New York, Vol. 6, pp. 373–385.

Johnston, R. W., Osmundsen, H., and Park, M. V. (1979) 'Associative properties of butyryl-coenzyme A synthetase from ox liver mitochondria', *Biochim. Biophys. Acta.*, **569**, 70–81.

Jones, K. M. (1959) 'The mechanisms of nicotinuric acid synthesis', *Biochem. J.*, **73**, 714–719.

Jones, M. E., Lipman, F., Hilz, H., and Lynen, F. (1953) 'On the enzymatic mechanism of coenzyme A acetylation with adenosine triphosphate and acetate', *J. Am. Chem. Soc.*, **75**, 3285–3286.

Kaihara, M., and Price, J. M. (1961) 'Quinaldylglycyltaurine: A urinary metabolite of quinaldic acid and kynurenic acid in the cat', *J. Biol. Chem.*, **236**, 508–511.

Kamienski, F. X., and Casida, J. E. (1970) 'Importance of demethylation in the metabolism *in vivo* and *in vitro* of methylenedioxyphenyl synergists and related compounds in mammals', *Biochem. Pharmacol.*, **19**, 91–112.

Kao, J., Jones, C. A., Fry, J. R., and Bridges, J. W. (1978) 'Species differences in the metabolism of benzoic acid by isolated hepatocytes and kidney tubule fragments', *Life Sci.*, **23**, 1221–1228.

Kaslander, J., Sijpesteijn, A. K., and Van der Kerk, G. J. M. (1962) 'On the transformation of the fungicide sodium dimethyl dithiocarbamate into its alanine derivative by plant tissues', *Biochim. Biophys. Acta*, **60**, 417–419.

Kellerman, G. M. (1958) 'Benzoyl adenylate and hippuryl adenylate: preparation, properties and relationship to the synthesis and transport of hippurate', *J. Biol. Chem.*, **231**, 427–443.

Kelley, D. S., and Potter, V. R. (1978) 'Regulation of amino acid transport systems by amino acid depletion and supplementation in monolayer cultures of rat hepatocytes', *J. Biol. Chem.*, **253**, 9009–9017.

Kielley, R. K., and Schneider, W. C. (1950) 'Synthesis of *p*-aminohippuric acid by mitochondria of mouse liver homogenates', *J. Biol. Chem.*, **185**, 869–880.

Killenberg, P. (1978) 'Measurement and subcellular distribution of choloyl-CoA synthetase and bile acid-CoA: amino acid N-acyltransferase activities in rat liver', *J. Lipid Res.*, **19**, 24–31.

Killenberg, P., and Jordan, J. T. (1978) 'Purification and characterisation of bile acid-CoA: amino acid *N*-acyltransferase from rat liver', *J. Biol. Chem.*, **253**, 1005–1010.

Killenberg, P. G., and Webster, L. T., Jr. (1980) 'Conjugation by peptide bond formation' in *Enzymatic Basis of Detoxification* (W. B. Jakoby, Ed.), Academic Press Inc., New York, Vol. II pp. 141–167.

Killenberg, P. G., Davidson, E. D., and Webster, L. T., Jr. (1971) 'Evidence for a medium-chain fatty acid: coenzyme A ligase (adenosine monophosphate) that activates salicylate', *Molecular Pharmacology*, **7**, 260–268.

Knaak, J. B., and Sullivan, L. J. (1968) 'Metabolism of 3,4-dichlorobenzyl *N*-methylcarbamade in the rat', *J. Agr. Fd. Chem.*, **16**, 454–459.

Lan, S. J., El-Hawey, A. M., Dean, A. V., and Schreiber, E. C. (1975) 'Metabolism of *p*-(cyclopropylcarbonyl) phenylacetic acid (SQ 20,650). Species differences', *Drug Metab. Dispos.*, **3**, 171–179.

Lau, E. P., Haley, B. E., and Barden, R. E. (1977) 'Photoaffinity labeling of acyl-coenzyme A glycine *N*-acyltransferase with *p*-azidobenzoyl-coenzyme A', *Biochemistry*, **16**, 2581–2585.

Le Cam, A., and Freychet, P. (1977) 'Neutral amino acid transport; characterization of the A and L systems in isolated rat hepatocytes', *J. Biol. Chem.*, **252**, 148–156.

Lethco, E. J., and Brouwer, E. A. (1966) 'The metabolism of napthaleneacetic acid-1-C^{14} in rats', *J. Agr. Fd. Chem.*, **14**, 532–535.

Leuthardt, F., and Nielsen, H. (1951) 'Recherches sur la synthèse biologique de l'acide hippurique', *Helv. Chim. Acta*, **34**, 1618–1621.

Levy, G. (1971) 'Drug biotransformation interactions in man: non-narcotic analgesics', *Ann. N.Y. Acad. Sci.*, **179**, 32–42.

Lippel, K., and Beattie, D. S. (1970) 'The submitochondrial distribution of acid:CoA ligase (AMP) and acid:CoA ligase (GDP) in rat liver mitochondria', *Biochim. Biophys. Acta*, **218**, 227–232.

Litwack, G., Ketterer, B., and Arias, I. M. (1971) 'Ligandin: a hepatic protein which binds steroids, bilirubin, carcinogens and a number of exogenous organic anions', *Nature*, **234**, 466–467.

Londesborough, J. C., and Webster, L. T., Jr. (1974) 'Fatty acyl-CoA synthetases' in *Enzymes*, (P. D. Boyer, Ed.), 3rd edn. Academic Press, New York, Vol. 10, pp. 469–488.

Mahler, H. R., Wakil, S. J., and Bock, R. M. (1953) 'Studies on fatty acid oxidation I. Enzymatic activation of fatty acids' *J. Biol. Chem.*, **204**, 453–468.

McGiven, J. D., and Klingenberg, M. (1971) 'Correlation between H^+ and anion movement in mitochondria and the key role of the phosphate carrier', *Eur. J. Biochem.*, **20**, 392–399.

McNamara, P. D., and Ozegovic, B. (1980) 'Membrane structure and transport systems', in *Principles of Metabolic Control in Mammalian Systems* (R. H. Hermann, R. M. Cohn, and P. D. McNamara, Eds.), Plenum Press, New York, pp. 373–437.

Meister, A., and Tate, S. S. (1976) 'Glutathione and related γ-glutamyl compounds: biosynthesis and utilization', *Ann. Rev. Biochem.*, **45**, 559–604.

Millburn, P. (1978) 'Biotransformation of xenobiotics in animals' in *Biochemical aspects of plant and animal coevolution* (J. B. Harborne, Ed.), Academic Press, London, pp. 35–73.

Moldave, K., and Meister, A. (1957) 'Synthesis of phenylacetylglutamine by human tissue', *J. Biol. Chem.*, **229**, 463–476.

More, J. E., Roberts, T. R., and Wright, A. N. (1978) 'Studies of the metabolism of 3-phenoxybenzoic acid in plants', *Pestic. Biochem. Physiol.*, **9**, 268–280.

Mumma, R. O., and Hamilton, R. H. (1976) 'Amino acid conjugates' in *Bound and Conjugated Pesticide Residues*, (D. Kaufman, G. Still, G. Paulson, and S. Bandal, Eds.), ACS Symposium Series 29, American Chemical Society, Washington, D.C. pp. 68–85.

Murphy, G. M., and Signer, E. (1974) 'Bile acid metabolism in infants and children', *Gut*, **15**, 151–163.

Nandi, D. L., Lucas, S. V., and Webster, L. T., Jr. (1979) 'Benzoyl-coenzyme A: glycine *N*-acyltransferase and phenylacetyl-coenzyme A: glycine *N*-acyltransferase from bovine liver mitochondria', *J. Biol. Chem.*, **254**, 7230–7237.

Ohkawa, H., Kaneko, H., and Miyamoto, J. (1977) 'Metabolism of permethrin in bean plants', *J. Pestic. Sci.*, **2**, 67–76.

Osmundsen, H., and Park, M. V. (1975) 'Bovine serum albumin activation of ox liver butyryl coenzyme A synthetase', *Biochem. Soc. Trans.*, **3**, 327–329.

Osmundsen, H., and Sherratt, H. S. A. (1975) 'A novel mechanism for inhibition of beta-oxidation by methylenecyclopropylacetyl-CoA, a metabolite of hypoglycin', *FEBS Lett.*, **55**, 38–41.

Park, M. V. (1978) 'Acyl-coenzyme A synthetases', *Biochem. Soc. Trans.*, **6**, 70–72.

Parkki, M. G. (1978) 'Induction of glycine conjugation in rat liver', in *Conjugation reactions in drug biotransformation.* (A. Aito, Ed.), Elsevier North-Holland, Biomedical Press, Amsterdam, p. 525.

Percy-Robb, I. W., and Boyd, G. S. (1973) 'The biosynthesis of bile acids', *Scot. Med. J.*, **18**, 166–174.

Peric-Golia, L., and Jones, R. S. (1962) 'Ornithocholanic acids-abnormal conjugates of bile acids', *Proc. Soc. Exp. Biol. Med.*, **110**, 327–331.

Peric-Golia, L., and Jones, R. S. (1963) 'Ornithocholanic acids and cholelithiasis in man', *Science*, **142**, 245–246.

Phillip, D. P., and Parsons, P. (1979a) 'Isolation and purification of long chain fatty acyl coenzyme A ligase from rat liver mitochondria', *J. Biol. Chem.*, **254**, 10776–10784.

Phillip, D. P., and Parsons, P. (1979b) 'Kinetic characterization of long chain fatty acyl coenzyme A ligase from rat liver mitochondria', *J. Biol. Chem.*, **254**, 10785–10790.

Pinto, J. F., Camien, M. N., and Dunn, M. S. (1965) 'Metabolic fate of p,p'-DDT[1,1,1,tri-chloro-2,2-bis (*p*-chlorophenyl)ethane] in rats', *J. Biol. Chem.*, **240**, 2148–2154.

Quick, A. J. (1931) 'The conjugation of benzoic acid in man', *J. Biol. Chem.*, **92**, 65–85.

Quick, A. J. (1932a) 'The site of the synthesis of hippuric acid and phenylacturic acid in the dog', *J. Biol. Chem.*, **96**, 73–81.

Quick, A. J. (1932b) 'The relationship between chemical structure and physiological response. I. The conjugation of substituted benzoic acids', *J. Biol. Chem.*, **96**, 83–101.

Renjeva, R., Boudet, A. M., and Faggion, R. (1976) 'Phenolic metabolism in petunia tissues. IV. Properties of *p*-coumarate: coenzyme A ligase isoenzymes' *Biochimie*, **58**, 1255–1262.

Roberts, T. R. (1981) 'The metabolism of the synthetic pyrethroids in plants and soils', in *Progress in Pesticide Biochemistry*, (D. H. Hutson and T. R. Roberts, Ed.), John Wiley, Chichester, Vol. 1, pp. 115–146.

Scandurra, R., Politi, L., Dupré, S., Moriggi, M., Barra, D., and Cavallini, D. (1977) 'Comparative biological production of taurine from free cysteine and from cysteine bound to phosphopantothenate', *Bull. Mol. Biol. Med.*, **2**, 172–177.

Scandurra, R., Federici, G., Dupré, S., and Cavallini, D. (1978) 'Taurine and isethionic acid production in mammals', *Bull. Mol. Biol. Med.*, **3**, 141–147.

Schachter, D., and Taggart, J. V. (1953) 'Benzoyl coenzyme A and hippurate synthesis', *J. Biol. Chem.*, **203**, 925–934.

Schachter, D., and Taggart, J. V. (1954) 'Glycine *N*-acylase: purification and properties', *J. Biol. Chem.*, **208**, 263–275.

Schersten, T. (1967) 'The synthesis of taurocholic and glycocholic acids by preparations of human liver. I. Distribution of activity between subcellular fractions', *Biochim. Biophys. Acta.*, **141**, 144–154.

Schwartz, D. P., and Pallansch, M. J. (1962) 'Occurrence of phenylacetylglutamine in cows' milk', *Nature*, **194**, 186.

Shirkey, R. J., Kao, J., Fry, J. R., and Bridges, J. W. (1979) 'A comparison of xenobiotic metabolism in cells isolated from rat liver and small intestinal mucosa', *Biochem. Pharmcol.*, **28**, 1461–1466.

Shono, T., Unai, T., and Casida, J. E. (1978) 'Metabolism of permethrin isomers in American cockroach adults, housefly adults and cabbage looper larvae', *Pestic. Biochem. Physiol.*, **9**, 96–106.

Siperstein, M. D., and Murray, A. W. (1955) 'Enzymatic synthesis of choloyl-CoA and taurocholic acid', *Science*, **138**, 377–378.

Skrede, S., and Bremer, J. (1970) 'The compartmentation of CoA and fatty acid activating enzymes in rat liver mitochondria', *Eur. J. Biochem.*, **14**, 465–472.

Skrede, S., and Halvorsen, O. (1979) 'Mitochondrial biosynthesis of coenzyme A', *Biochem. Biophys. Res. Commun.*, **91**, 1536–1542.

Sottocasa, G. L., Kuylenstierna, B., Ernster, L., and Bergstrand, A. (1967) 'Separation and some enzymatic properties of the inner and outer membranes of rat liver mitochondria' *Methods in Enzymology.*, **10**, 448–463.

Stern, K., Tyhurst, J. S., and Askonas, B. A. (1946) 'Note on hippuric acid synthesis in senility', *Amer. J. Med. Sci.*, **212**, 302–305.

Strahl, N. R., and Barr, W. H. (1971) 'Intestinal drug absorption and metabolism. 3. Glycine conjugation and accumulation of benzoic acid in rat intestinal tissue', *J. Pharm. Sci.*, **60**, 278–281.

Strange, R. C., Cramb, R., Hayes, J. D., and Percy-Robb, I. W. (1977) 'Partial purification of two lithocholic acid-binding proteins from rat liver 100,000 g supernatants', *Biochem. J.*, **165**, 425–429.

Strange, R. C., Chapman, B. T., Johnston, J. D., Nimmo, I. A., and Percy-Robb, I. W. (1979) 'Partitioning of bile acids into subcellular organelles and the *in vivo* distribution of bile acids in rat liver', *Biochim. Biophys. Acta.*, **573**, 535–545.

Thompson, G. A., and Meister, A. (1980) 'Modulation of γ-glutamyl transpeptidase activities by hippurate and related compounds', *J. Biol. Chem.*, **255**, 2109–2113.

Tishler, S. L., and Goldman, P. (1970) 'Properties and reactions of salicyl-coenzyme-A', *Biochem. Pharmacol.*, **19**, 143–150.

Van den Oord, A., Danielsson, H., and Ryhage, R. (1965) 'On the structure of the emulsifiers in gastric juice from the crab, *Cancer pagurus L.*', *J. Biol. Chem.*, **240**, 2242–2247.

Venis, M. A. (1972) 'Auxin-induced conjugation systems in peas', *Plant Physiol.*, **49**, 24–27.

Vessey, D. A. (1978) 'The biochemical basis for the conjugation of bile acids with either glycine or taurine', *Biochem. J.*, **174**, 621–626.

Vessey, D. A. (1979) 'The copurification and common identity of choloyl CoA:glycine- and choloyl CoA:taurine-*N*-acyltransferase activities from bovine liver', *J. Biol. Chem.*, **254**, 2059–2063.

Vessey, D. A., Crissey, M. H., and Zakim, D. (1977) 'Kinetic studies on the enzymes conjugating bile acids with taurine and glycine in bovine liver', *Biochem. J.*, **163**, 181–183.

Vessey, D. A., and Zakim, D. (1977) 'Characterization of microsomal cholyl coenzyme A synthetase', *Biochem. J.*, **163**, 357–362.

Vest, M. F., and Salzberg, R. (1965) 'Conjugation reactions in the newborn infant: the metabolism of para-aminobenzoic acid', *Arc. Dis. Childh.*, **40**, 97–105.

Wallcave, L., Bronczyk, S., and Gingell, R. (1974) 'Excreted metabolites of 1,1,1-trichloro-2,2-bis(*p*-chlorophenyl) ethane in the mouse and hamster', *J. Agr. Fd. Chem.*, **22**, 904–908.

Walton, E., and Butt, V. S. (1971) 'The demonstration of cinnamyl-CoA synthetase activity in leaf extracts', *Phytochemistry*, **10**, 295–304.

Wan, S. H., and Riegelman, S. (1972) 'Renal contribution to overall metabolism of drugs. II. Biotransformation of salicyclic acid to salicyluric acid', *J. Pharm. Sci.*, **61**, 1284–1287.

Webster, L. T., Jr. (1963) 'Studies of the acetyl coenzyme A synthetase reaction 1. Isolation and characterization of enzyme-bound acetyl adenylate' *J. Biol. Chem.*, **238**, 4010–4015.

Webster, L. T., Jr., and Campagnari, F. (1962) 'The biosynthesis of acetyl and butyryl adenylates', *J. Biol. Chem.*, **237**, 1050–1055.

Webster, L. T., Jr., Gerowin, L. D., and Rakita, L. (1965) 'Purification and characteristics of a butyryl coenzyme A synthetase from bovine heart mitochondria', *J. Biol. Chem.*, **240**, 29–33.

Webster, L. T., Jr., Mieyal, J. J., and Siddiqui, U. A. (1974) 'Benzoyl and hydroxybenzoyl esters of coenzyme A: ultraviolet characterization and reaction mechanisms', *J. Biol. Chem.*, **249**, 2641–2645.

Webster, L. T., Jr., Siddiqui, U. A., Lucas, S. V., Strong, J. M., and Mieyal, J. J. (1976) 'Identification of separate acyl-CoA:glycine and acyl-CoA: L-glutamine *N*-acyltransferase activities in mitochondrial fractions from liver of rhesus monkey and man', *J. Biol. Chem.*, **251**, 3352–3358.

Wright, A. N., Roberts, T. R., Dutton, A. J., and Doig, M. V. (1980) 'The metabolism of cypermethrin in plants: The conjugation of the cyclopropyl moiety', *Pestic. Biochem. Physiol.*, **13**, 71–80.

Yang, R. S. H. (1976) 'Enzymatic conjugation and insecticide metabolism', in *Insecticide Biochemistry and Physiology*, (C. F. Wilkinson, Ed.), Heyden, London, pp. 177–225.

Yousef, I. M., and Fisher, M. M. (1975) 'Bile acid metabolism in mammals. VIII. Biliary secretion of cholylarginine by the isolated perfused rat liver', *Can. J. Physiol. Pharmacol.*, **53**, 880–887.

Progress in Pesticide Biochemistry, Volume 2
Edited by D. H. Hutson and T. R. Roberts
© 1982 John Wiley & Sons, Ltd

CHAPTER 5

Formation of lipophilic conjugates of pesticides and other xenobiotic compounds

D. H. Hutson

INTRODUCTION

The metabolism of xenobiotic organic compounds usually involves a multi-step sequence of enzyme-catalysed biotransformations. These have been broadly categorized into primary and secondary phases. Primary biotransformations either introduce or expose functional groups which may then participate in other processes. Two classical examples of primary reactions are (i) the introduction, by oxidation, of a hydroxyl group into chemically inert alkanes and aromatic hydrocarbons and (ii) the exposure of carboxylic acid and alcohol functions by the hydrolysis of esters such as the pyrethroid insecticides. Secondary biotransformations are those in which these functional groups are linked to endogenous molecules to form conjugates. Conjugation reactions are

Abbreviations

The shorthand notation 16:0, 18:0, 18:1, 18:2, etc. used in the text represents chain length: number of double bonds.

enzyme-catalysed group transfer reactions in which, therefore, there is usually a donor–acceptor relationship. Three types of mechanism are involved:

1. Reaction of the xenobiotic (or a primary metabolite thereof) with an active endogenous donor, e.g. the reaction of 12-hydroxydieldrin with uridine-diphosphoglucuronic acid (UDPGA) to afford 12-hydroxydieldrin glucuronide.
2. Activation of the xenobiotic to form a donor molecule which can then react with an endogenous acceptor, e.g. the activation of 2,4-dichlorophenoxyacetic acid to its coenzyme A derivative which undergoes acyl transfer to an amino acid such as glycine.
3. Reaction of xenobiotic electrophiles with the most readily available endogenous soft nucleophile, glutathione. The resulting glutathione conjugates are further metabolized by reactions which differ in sequence and final products depending on whether they occur in mammals, plants or insects.

The majority of conjugation reactions involve the linkage of xenobiotics with sugars (glucose, glucuronic acid, etc), sulphate ion, a variety of amino acids, glutathione and acetate. In most cases, because the physical properties of the original molecule have been so drastically altered, its original bioactivity is destroyed. It must be stressed, however, that there are some dramatic exceptions to this generalization and it is quite possible that conjugation confers a change in type of bioactivity and, even in some cases, may be responsible for bioactivity, toxicity, etc (see Chapter 6). The classical conjugates are also much more water-soluble than their precursors and, in addition, often participate in active secretion processes, particularly from mammalian organs (e.g. from liver into bile). Conjugation thus facilitates elimination of xenobiotics from the body.

Recently, however, improved methods of isolation, purification and identification of metabolites have led to the characterization of a number of xenobiotic metabolites which are not rapidly eliminated. These metabolites are lipid-like in character. While their physical properties distinguish them from the classical conjugates, their mechanisms of formation clearly define them as conjugates, i.e. they are formed by the linking of xenobiotic metabolites to endogenous molecules. A variety of these compounds, lipophilic conjugates, are described below and their significance to mode of action and toxicity is assessed.

GLYCERYL ESTERS OF ALIPHATIC ACIDS

Xenobiotic fatty acids

Cycloprate

The miticidally active hexadecyl cyclopropanecarboxylate (cycloprate, Figure 1a), when dosed to mammals, is hydrolysed to afford cyclopropanecarboxylic acid and hexadecyl alcohol. The acid undergoes a chain-lengthening process to

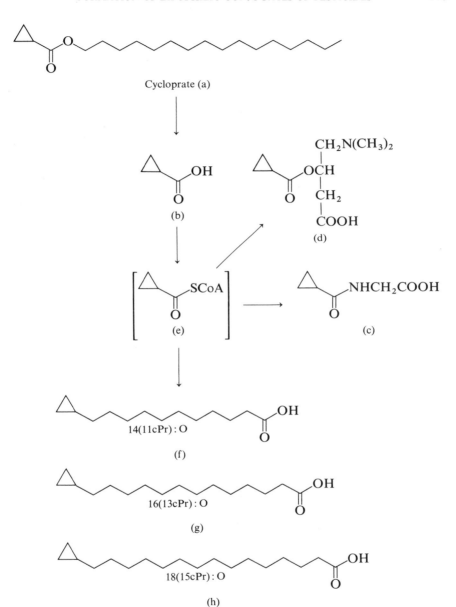

Figure 1 Metabolism of cycloprate in mammals

form several cyclopropylalkanoic acids (Figure 1) which are able to participate in the normal pathways of triglyceride synthesis. The over-all process is quite efficient; thus in rats (Quistad et al., 1978a) although about 82 % of the carbon-14 of an oral dose (21 mg kg^{-1}) of [^{14}C-*carboxyl*]cycloprate was eliminated in the urine and faeces between 0 h and 96 h after treatment, the remainder was retained in the carcass. At 15 days, 10 % of the administered dose remained in the animals. Most of the urinary metabolite was N-(cyclopropylcarbonyl) glycine (Figure 1c) but, significantly, 1–2 % was identified as N-(cyclopropylcarbonyl)carnitine (Figure 1d). The fat contained the highest radiochemical residue at four days but only 3 % of this was due to unchanged cycloprate; almost 90 % was found in the triglyceride fraction. Muscle, liver and carcass (minus the major organs) also contained radioactive triglycerides. When these fractions were analysed by reverse-phase liquid chromatography (after saponification and methylation of the resultant fatty acids to their methyl esters) a mixture of cyclopropyl-alkanoic acids (Figure 1f,g,h) could be detected. The 16(13cPr):0 acid (Figure 1g) predominated over the other analogues in each case.

The radioactivity was cleared from the carcass in an approximately biphasic process with first and second half-lives of ten and 42 days, respectively. This turnover rate is similar to that of the natural triglycerides.

The fate of cycloprate in cows (Quistad et al., 1978b) was similar; e.g. residual radioactivity in milk was mostly due to xenobiotic triacylglycerols but, at the lower doses used (0.3 mg. kg^{-1}), the residue levels were much lower than those in the rat, and elimination was virtually quantitative after six to seven days. Interestingly, the situation in dogs (Quistad et al., 1978c) was dramatically different. Following a dose of 0.7 mg. kg^{-1}, 73–77 % of the radioactivity was found to be retained 96 h after treatment. This was mostly in muscle (56 %) but with appreciable concentrations in kidney (7 %), liver (2 %), fat (6 %) and skin (6 %). Half of the residue in muscle was found to be the carnitine conjugate (Figure 1d). That in adipose tissue was mainly present as a mixture of triacyl-glycerols with the 16(13cPr):0 acid as the predominating xenobiotic acid. A substantial amount of unchanged cycloprate was also found in this tissue. The three acids (Figure 1f,g,h) were also the main xenobiotic components of the liver and kidney residue, but in these organs 28 % and 67 % of the radioactivity, respectively, was present as phospholipid. The major excretion product was the carnitine conjugate (Figure 1d), as opposed to the glycine conjugate in rat and cow. Because of the retention of the carnitine conjugate in dog muscle, the over-all rate of elimination of the pesticide from this species would be slow.

The chain-elongation process was also shown to occur in the foliage and fruit of apple and orange (Quistad et al., 1978d) but the cyclopropyl fatty acids were found as polar conjugates rather than as lipids in these plants. Unsaturated fatty acids were also detected.

These findings represent some pioneering work in that, not only was the role of carnitine in xenobiotic metabolism discovered, but the results suggest that a reinvestigation of the metabolism of other carboxylic acid metabolites of pesticides, etc, would be worthwhile. In addition, the ambitious attempts to analyse the array of radioactive lipophilic materials (which included some cholesteryl esters) was praiseworthy.

In considering the significance of these results in terms of safety assessment, the role of carnitine overshadows the phenomenon of the formation and storage of xenobiotic triacylglycerols. Currently there is no evidence to show that the latter process constitutes a toxicological problem. On the other hand, carnitine plays a vital role in the normal utilization of depot fat (long-chain fatty acids) as an energy source. The impairment of this role results in an accumulation of triacylglycerols in carnitine-deficient tissues (Bressler, 1970). The authors of these reports consider that the greater sensitivity of dogs versus rats in subacute toxicity studies with cycloprate is due to the immobilizing of carnitine as the conjugate (Figure 1d) in the muscle of dog.

Dodecylcyclohexane

Dodecylcyclohexane (chosen as a typical naphthenic constituent of mineral oils) was tritiated and fed (20 mg and 200 mg) to rats in their diet (Tulliez and Bories, 1979). The compound underwent extensive ω-oxidation to afford cyclohexyldodecanoic acid which was incorporated into both neutral lipids and phospholipids in liver and fat. The molecule was generally tritium labelled (Wilzbach method) so much of the label may have become incorporated into natural constituents. However, gas-liquid radiochromatography confirmed the presence of cyclohexyldodecanoic and cyclohexyldecanoic acids (as methyl ester) which were isolated from the neutral lipids of liver.

Other examples of chain elongation

An early example of chain elongation in pesticide chemistry was that of 4-(2,4-dichlorophenoxy)butyric acid (1) to 2,4-dichlorophenoxyhexanoic acid (2) and higher homologues in alfalfa plants (Linscott et al., 1968). Herbicide (1) owes its action to β-oxidation in sensitive species. It is proposed that, as well as only slowly β-oxidizing the compound, alfalfa and other legumes may 'detoxify'

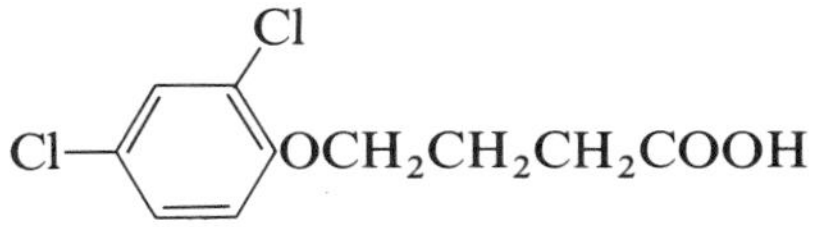

Cl

Cl—⟨benzene ring⟩OCH$_2$CH$_2$CH$_2$COOH

(1)

⟨benzene ring⟩—OCH$_2$CH$_2$CH$_2$CH$_2$CH$_2$COOH

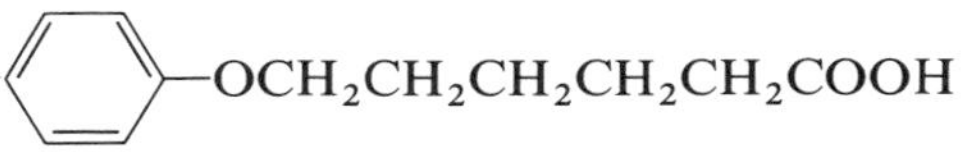

(2)

(1) by the chain-elongation process. Similarly, 2,4-dichlorophenoxyacetic acid (2,4-D), applied as its methyl ester, to alfalfa was partially converted into the butanoic and hexanoic acid analogues (Linscott and Hagin, 1970). The hypotensive agent and dopamine β-hydroxylase inhibitor, 5-(4-chlorobut-1-yl)-picolinic acid (CBPA, 3), was converted into five metabolites when orally administered to rats (Miyazaki et al., 1976) four of which were elongated by a C_2 unit to afford, for example, metabolite (4).

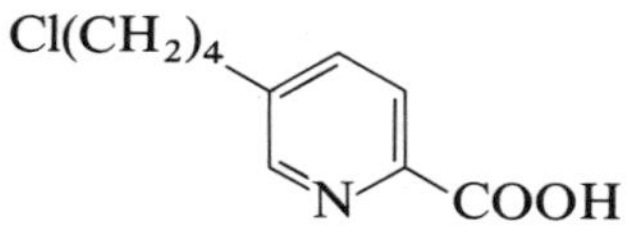

(3)	(4)

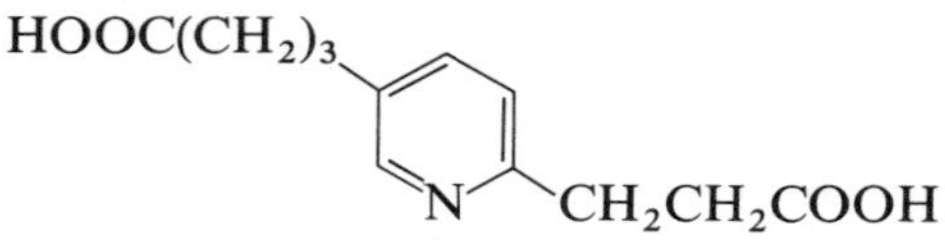

There is no evidence for the incorporation of elongated CBPA or elongated 2,4-dichlorophenoxyacetic acid homologues into lipids. Whether or not this happens presumably depends on the requirement for a specific length of alkane chain below which the substrate is regarded by lipid metabolism systems as 'too xenobiotic' for incorporation. However, as discussed below, aromatic acids can be directly incorporated into triglycerides.

Other aliphatic acids

A study of the metabolism of [5-^{14}C]methoprene [isopropyl (2E,4E)-11-methoxy-3,7,11-trimethyl-2,4-dodecadienoate, Figure 2a] in Leghorn chickens has revealed minor metabolites conjugates as xenobiotic triglycerides (Quistad et al., 1976). Extensive labelling of natural components occurred owing to the incorporation of part of the aliphatic chain into normal biochemical processes. However, thorough analysis of fat, liver and egg yolks revealed that methoprene acid (Figure 2b), after saturation to 11-methoxy-3,7,11-trimethyldodecanoic acid (Figure 2c, MTDA) and partial demethylation to its 11-hydroxy analogue (HTDA), was conjugated in the form of triglycerides. MTDA glyceride accounted for 31 % of the radioactivity in egg yolk following oral dosing at 77 mg kg^{-1}. (This dropped to 2 %, 0.06 mg kg^{-1}, at a realistic dose of 0.60 mg kg^{-1}.) HTDA glyceride accounted for only 2 % and 0.4 % at these respective doses. Residues in fat, muscle and liver were much lower than this. The precise nature of the xenobiotic triglycerides was not determined but 1,3-distearoyl and 1,2-dioleyl-MTDA-glycerols were used as reference compounds. Diglycerides and cholesteryl esters (see below) were also found. The analogous phytanic acid (5) and its homologues, which are widely

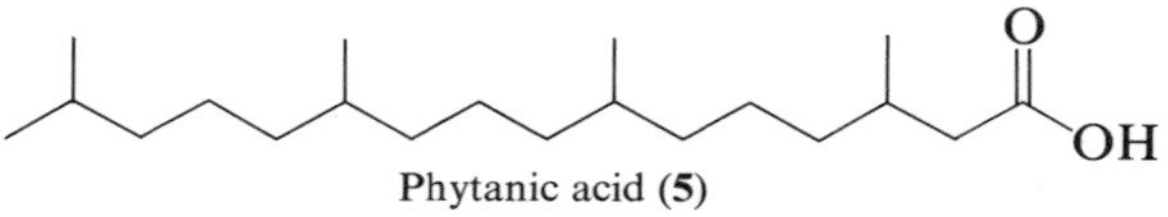

Phytanic acid (5)

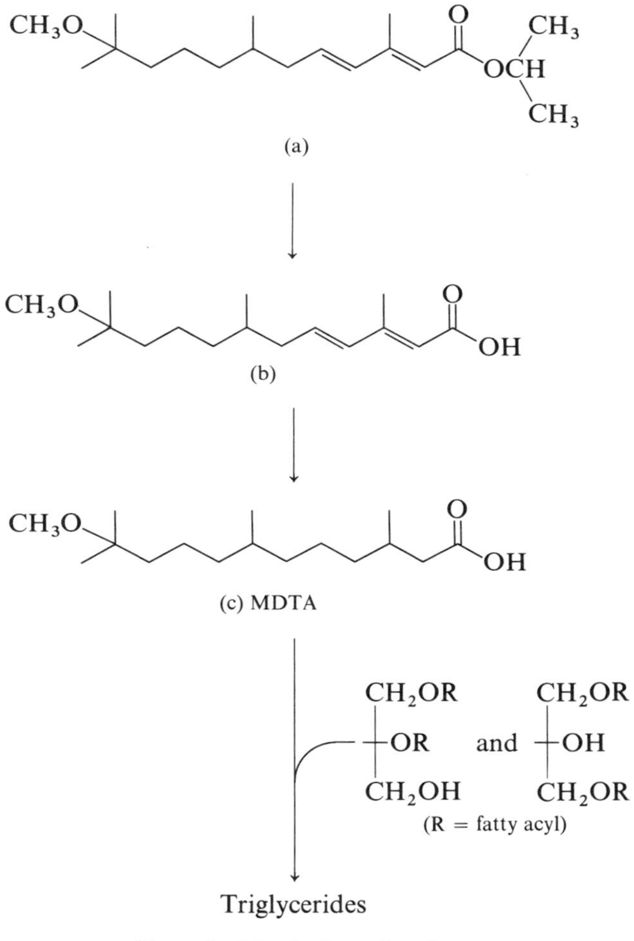

Figure 2 Metabolism of methoprene

distributed in nature, are also found incorporated into triglycerides and other types of lipid (Lough, 1973). Indeed, there exists for humans a rare nervous disorder, Refsums disease (*Heredopathia atactica polyneuritiformis*) in which dietary phytanic acid, etc becomes incorporated in high concentrations into tissue lipids (Klenk and Kahlke, 1963). The accumulation of phytanic acid in the first instance is thought to be due to a metabolic defect in its catabolism. However, these considerations are of only academic importance in relation to compounds like methoprene since, for the individuals in whom Refsums disease is manifest, dietary sources of these isoprenoid acids would far outweigh any derived from methoprene.

GLYCERYL ESTERS OF AROMATIC ACIDS

3-Phenoxybenzoic acid

3-Phenoxybenzoic acid (Figure 3a), a metabolite common to the important new synthetic pyrethroid insecticides permethrin, cypermethrin, decamethrin and fenvalerate (Hutson, 1979), is characterized by remarkable species differences in its conjugation in animals. It is conjugated with glycine (Figure 3c) in dogs and ferrets, glutamic acid in cows, taurine in mice, glycylvaline in mallard ducks and with glucuronic acid in marmosets (Huckle *et al.*, 1981; and see Chapter 4). In plants, it is conjugated with sugars, including disaccharides (see Chapter 3). During studies of its metabolism in rats, a small proportion $(1.3–3.0\%)$ of an oral dose $(0.76 \text{ mg kg}^{-1})$ was found in the skin four days after dosing. A similar proportion was found after daily dosing at 100 mg kg^{-1} (Crayford and Hutson, 1980). Part of this residue ($60–70\%$ at the high dose and 9% at the low dose) was incorporated into neutral lipids. Only 10% of the carcass residue was lipid, the remainder was unchanged 3-phenoxybenzoic acid. The skin was extracted and fractionated using gel-permeation, thin-layer and high-performance liquid chromatography with a view to isolating the (radioactive) metabolite in pure form. Mass spectral analysis indicated that the metabolite was 3-phenoxybenzoyldipalmitin (3,d,3-phenoxybenzoyldipalmitoyl-glycerol). The positions of substitution could not be deduced from the mass spectrum but comparison with authentic standards indicated that both the 1- and 2-positional isomers had been formed. The discovery of 3-phenoxy-benzoyldipalmitin suggests a degree of specificity in the incorporation of the xenobiotic acid into the triglycerides. However, in taking the approach of rigorous purification, undoubtedly several other triglycerides were removed

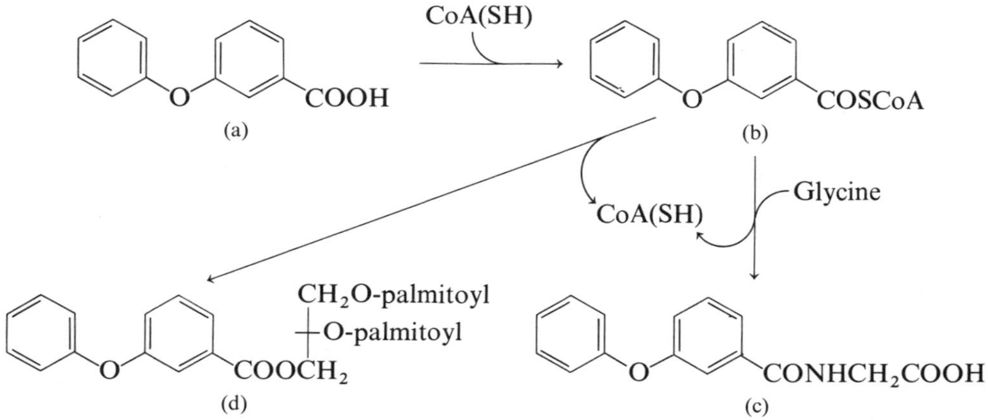

Figure 3 Conjugation of 3-phenoxybenzoic acid with amino acid and diglyceride

from the main fraction and lost in the process. The reaction is a very minor one at low doses and when dosing esters (e.g. cypermethrin or 3-phenoxybenzoyl glucoside). The acid is the major residue under these conditions, though it is interesting to speculate that it may have reached the skin by transport as a triglyceride. Such a mechanism would explain why an acid, fully ionized at physiological pH, was found unexpectedly in the skin. The extensive amino acid conjugation undergone by 3-phenoxybenzoic acid (see Chapter 4) occurs via the formation of 3-phenoxybenzoyl-coenzyme A (Figure 3b). It is very likely that acyl transfer to glycerides is an alternative reaction to amino acid conjugation (Figure 3). Because these aroyl glycerides are esters, they are likely to be substrates for esterase and lipase action and will not be stored indefinitely but, as with their aliphatic analogues, will be turned over at a rate similar to that of natural fats.

Ethyl 4-benzyloxybenzoate

Preceding the above pesticide-related discovery of aromatic acid involvement in xenobiotic lipid formation, Fears *et al.* (1978) reported the phenomenon with the hypolipidemic agent, ethyl 4-benzyloxybenzoate (**6**). The reaction was discovered *in vitro* and confirmed *in vivo* by dosing rats with large amounts of (**6**) (5000 mg kg^{-1}). Rigorous purification of the xenobiotics from epididymal adipose tissue afforded a mixture of three metabolites which could be identified by mass spectral analysis (high resolution measurements, etc) as:

4-benzyloxybenzoyl dipalmitoyl glycerol,
4-benzyloxybenzoyl oleoyl palmitoyl glycerol, and
4-benzyloxybenzoyl myristoyl palmitoyl glycerol.

The *in vitro* system, involving rat liver slices and intestinal rings, the xenobiotic acid and [^{14}C]glycerol, proved useful in that a large number of acids could be screened for incorporation; 15 acids were so screened. Several were found not to be incorporated (e.g. nicotinic acid and acetylsalicylic acid), four (**7**, **8**, **9** and **10**) were incorporated better than (**6**), while the similarly sized (**11**) was very poorly incorporated. Although the significance of the process is not yet fully understood, the demonstration of the involvement of these xenobiotic acids in lipid metabolism may throw light on the mechanism of their pharmacological action.

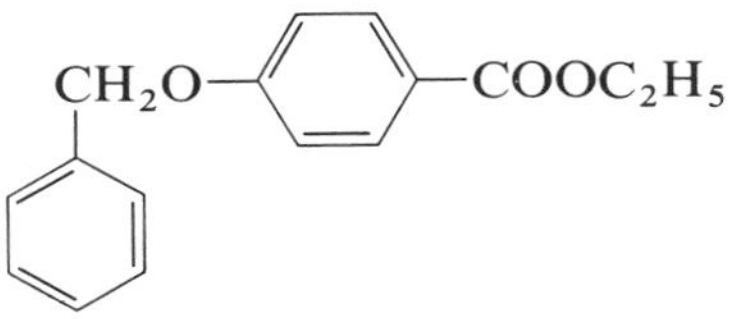

(6) 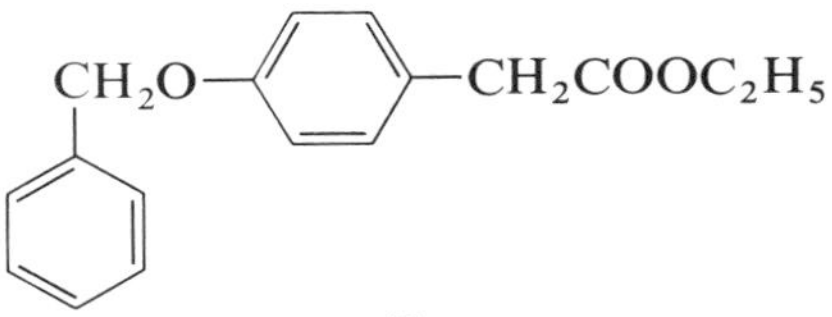   (7)

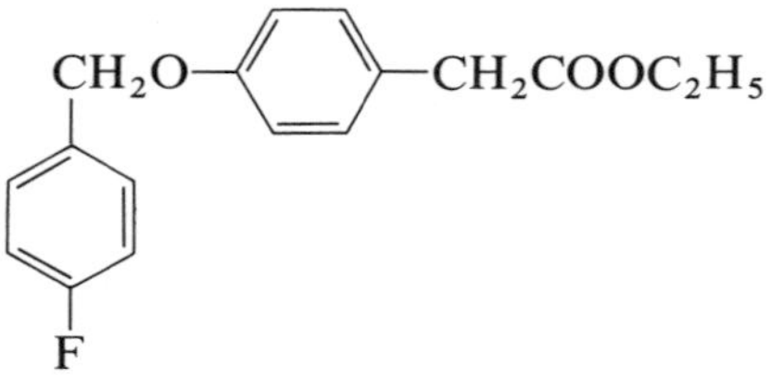

(8)

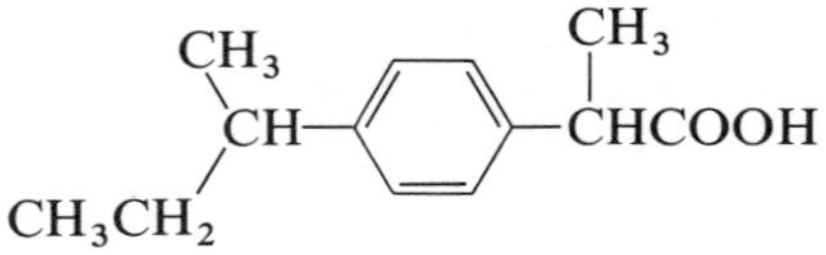

Ibuprofen (9)

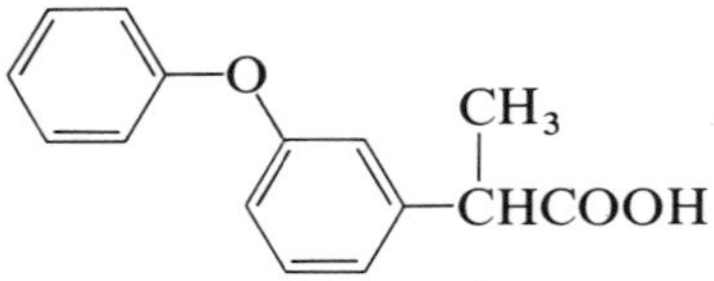

Fenoprofen (10) 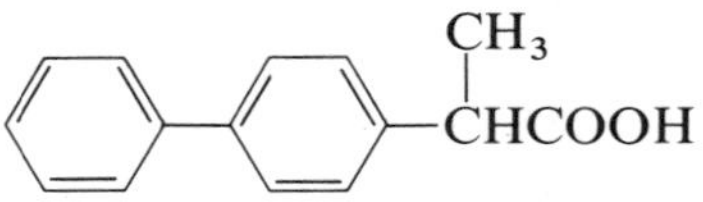   Flurbiprofen (11)

Strictly speaking, compounds (7)–(11) are aliphatic acids/esters and should be included in the previous section. However, it seems likely that lipophilic character is more important than the aromatic or aliphatic nature of the substituent on the carboxylic acid group. It seems likely that short-chain aliphatic acids and single-ring aromatic acids (e.g. benzoic and acetylsalicylic acid) are not lipophilic enough to participate in these reactions.

ESTERS OF CHOLESTEROL

The experiments reported above on the insect growth regulant, methoprene (Figure 2a), also revealed the presence of cholesteryl MTDA (12) as the major residue (22 μg g^{-1}) in the livers of chickens dosed with [^{14}C]methoprene (64 mg kg^{-1}) (0.0024 μg g^{-1} at 0.6 mg kg^{-1}). The metabolism of cholesterol in the liver is complex and involves uptake, synthesis, utilization for bile acid and lipoprotein production, as well as esterification to form cholesteryl esters. It is not known whether the cholesteryl esters function just as a storage form for cholesterol or whether they are incorporated into newly synthesized lipoprotein (Lichtenstein and Brecher, 1980). With these uncertainties, it is not possible to assess the significance of the formation of xenobiotic cholesteryl esters except

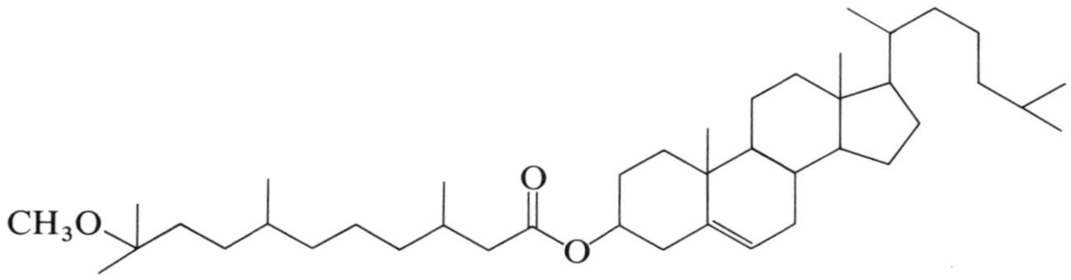

(12)

to say that they are probably exported from the liver, widely distributed and subjected to reasonably rapid catabolism.

The conjugation processes outlined so far in this chapter are examples of mechanism (2) described in the introduction.

FATTY ACID ESTERS OF XENOBIOTIC ALCOHOLS

DDT

Following a series of studies in which the conjugation of cannabis-related alcohols with normal fatty acids was discovered (see below), Leighty and co-workers (1980) have found that such conjugation reactions occur with a metabolite of DDT, DDOH (13). Rats were dosed intraperitoneally for four days with DDT (200 mg kg^{-1}) and, seven days after the first dose, their livers and spleens were extracted with chloroform. Synthetic palmitoyl-DDOH (14),

(13)   (14)

1-palmitoyloxy-2,2-bis-(p-chlorophenyl)ethane was used as a reference compound. EI and CI mass spectrometry of the t.l.c. and g.c. purified metabolites characterized them as the DDOH conjugates of palmitic (16:0) stearic (18:0), oleic (18:1) and linoleic (18:2) fatty acids. These metabolites could also be biosynthesized *in vitro* from DDOH, rat liver microsomes and a CoA/ATP/Mg^{2+} cofactor mixture. It is postulated that fatty acid conjugation of DDT may be a mechanism by which some of its metabolites are retained in the body. However, the rather long retention of DDT itself and the very long retention of its metabolite DDE, clearly unrelated to lipoconjugation, are probably more important.

Cannabinoids

When ^{14}C-labelled Δ^8- or Δ^9-tetrahydrocannabinol (15 and 16) is administered to rats, retention of radioactivity in liver, spleen and fat is observed (Leighty, 1973). This has been shown to be due to the conjugation of the 11-hydroxylated analogues, (17) and (18), with endogenous fatty acids (Leighty et al., 1976). Palmitic and stearic acids are mainly utilized in this process, with oleic and linoleic acid conjugates present in lesser quantities. The metabolites could be biosynthesized using a rat liver microsomal system, cofactors (as for DDOH above) and the 11-hydroxytetrahydrocannabinols (17) and (18) as acceptor substrates (Leighty, 1979).

(* Site of conjugation with fatty acid)

The significance of the retention of cannabis-related metabolites by fatty acid conjugation is uncertain. As 11-hydroxy-Δ^9-tetrahydro-cannabinol possesses physchomimetic activity, it has been suggested (Leighty et al., 1976) that such a storage mechanism, coupled with esterase-catalysed hydrolysis, could account for the 'flashbacks' associated with the use of marihuana.

Conjugations of xenobiotic alcohols with endogenous fatty acids exemplify mechanism (1) described in the introduction, i.e. reaction of the xenobiotic with a biochemically reactive endogenous donor (fatty acid CoA derivative). Similar metabolites, but of cannabinol, have been isolated from the faeces of rats (Yisak et al., 1978). The enzyme system catalysing the acylation of the cannabis-related alcohols may be the same as that catalysing the synthesis of cholesterol esters (Leighty, 1980).

CONCLUSIONS

This short chapter collects ample evidence to show that xenobiotic organic chemicals can participate in lipid biosynthetic pathways just as they do in other biochemical processes. The consequences of the formation of neutral xenobiotic lipids are several-fold. Unlike other conjugation modes, these processes delay the excretion of xenobiotics. At a constant daily intake, it is conceivable that residues of xenobiotic lipids could rise to appreciable levels. The consequences to the organism depend on the type and amount of bioactivity retained by the metabolite before (and after) conjugation. In any event, classical long-term toxicology studies in mammals should adequately cover such eventualities. Most evidence so far suggests that, at low dietary intakes of xenobiotics, xenobiotic lipids form only very low residues. Their ester-like nature should render them labile to carboxylesterases and lipases and it is probable, as has been shown with cycloprate, that these lipids have a turnover time similar to that of their natural counterparts.

Of particular interest in relation to the xenobiotic glycerides, is the possible formation of xenobiotic phospholipids. These may become involved in the formation of faulty membranes and possibly have effects on the nervous system. 3-Phenoxybenzoic acid has been tested in rats by observational and bio-chemical toxicology and no such effects found (A. J. Dewar and G. P. Rose, unpublished work, 1981). Presumably, the numerous hypolipidemic drugs (e.g. **6**, **9**, **10** and **11**) have also been tested for such neurochemical effects.

On balance, it would seem likely that, at realistic doses and application rates, xenobiotic lipid formation will not afford any toxicological problems. However, as more thorough metabolism studies are carried out with improved technology, more examples of these types of conjugates will be discovered.

NOTE ADDED IN PROOF

A recent report describes the major residue of carbofuran in carrots as the conjugate of 3-hydroxycarbofuran with (Z)-2-methyl-2-butenoic acid (angelic acid). (Sonobe, H., Kamps, L. R., Mazzola, E. P., and Roach, J. A. G. *J. Agr. Fd. Chem.* **29**, 1125–1129 (1981)).

REFERENCES

Bressler, R. (1970) in *Lipid Metabolism* (S. J. Wakil, Ed.), Academic Press, New York, p. 49.

Crayford, J. V., and Hutson, D. H. (1980) 'Xenobiotic triglyceride formation', *Xenobiotica*, **10**, 349–354.

Fears, R., Baggaley, K. H., Alexander, R., Morgan, B., and Hindley, R. M. (1978) 'The participation of ethyl 4-benzyloxybenzoate (BRL 10894) and other aryl-substituted acids in glycerolipid metabolism', *J. Lipid Res.*, **19**, 3–11.

Huckle, K. R., Hutson, D. H., and Millburn, P. (1981) 'Species differences in the metabolism of 3-phenoxybenzoic acid', *Drug Metab. Disposit.*, **9**, 352–359.

Hutson, D. H. (1979) 'The metabolic fate of synthetic pyrethroid insecticides in mammals', in *Progress in Drug Metabolism* (J. W. Bridges and L. F. Chasseaud, Eds.), John Wiley, Chichester, Vol. 3, pp. 215–252.

Klenk, E., and Kahlke, W. (1963) 'Uber das Vorkommen der 3.7.11.15-Tetramethyl-hexadecansaure (Phytansaure) in den Cholinestern und anderen Lipoidfraktionen der Organe bei einem Krankheitsfall unbekanneter Genese (Verdacht auf Heredopathia atactica polyneuritiformis[Refsum-Syndrom])', *Z. Physiol. Chem.*, **333**, 133–139.

Leighty, E. G. (1973) 'Metabolism and distribution of cannabinoids in rats after different methods of administration', *Biochem. Pharmacol.*, **22**, 1613–1621.

Leighty, E. G. (1979) 'An *in vitro* rat liver microsomal system for conjugating fatty acids to 11-hydroxy-Δ^9-tetrahydrocannabinol', *Res. Commun. Chem. Pathol. Pharmacol.*, **23**, 483–492.

Leighty, E. G. (1980) 'β-Diethylaminoethyldiphenyl-propylacetate (SKF-525 A) and 2,4-dichloro-6-phenylphenoxyethylamine. HBr (DPEA) inhibition of fatty-acid conjugation to 11-hydroxy-Δ^9-tetrahydrocannabinol by the rat liver microsomal system', *Biochem. Pharmacol.*, **29**, 1071–1073.

Leighty, E. G., Fentiman, A. F., and Foltz, R. L. (1976) 'Long- retained metabolites of Δ^9- and Δ^8-tetrahydrocannabinols identified as novel fatty acid conjugates', *Res. Commun. Chem. Pathol. Pharmacol.*, **14**, 13–28.

Leighty, E. G., Fentiman, A. F., and Thompson, R. M. (1980) 'Conjugation of fatty acids to DDT in the rat: possible mechanisms for retention', *Toxicology*, **15**, 77–82.

Lichtenstein, A. H., and Brecher, P. (1980) 'Properties of acyl-CoA: cholesterol acyltransferase in rat liver microsomes', *J. Biol. Chem.*, **255**, 9098–9104.

Linscott, D. L., and Hagin, R. D. (1970) 'Additions to the aliphatic moiety of chlorophenoxy compounds', *Weed Sci.*, **18**, 197–198.

Linscott, D. L., Hagin, R. D., and Dawson, J. E. (1968) 'Conversion of 4-(2,4-dichlorophenoxy)butyric acid to homologues by alfalfa', *J. Agric. Food Chem.*, **16**, 844–848.

Lough, A. K. (1973) 'The chemistry and biochemistry of phytanic, pristanic and related acids', *Prog. Chem. Fats and Other Lipids*, **15, (part 1)**, 5–48.

Miyazaki, H., Takayama, H., Minatogawa, Y., and Miyano, K. (1976) 'A novel metabolic pathway in the metabolism of 5-(4'-chloro-*n*-butyl)picolinic acid', *Biomed. Mass Spectrom.* **3**, 140–145.

Quistad, G. B., Staiger, L. E., and Schooley, D. A. (1976) 'Environmental degradation of the insect growth regulator methoprene. X. Chicken metabolism', *J. Agric. Food Chem.*, **24**, 644–648.

Quistad, G. B., Staiger, L. E., and Schooley, D. A. (1978a) 'Environmental degradation of the miticide cycloprate (hexadecyl cyclopropanecarboxylate). 1. Rat metabolism', *J. Agric. Food Chem.*, **26**, 60–66.

Quistad, G. B., Staiger, L. E., and Schooley, D. A. (1978b) 'Environmental degradation of the miticide cycloprate (cyclohexyl cyclopropanecarboxylate). 3. Bovine metabolism', *J. Agric. Food Chem.*, **26**, 71–75.

Quistad, G. B., Staiger, L. E., and Schooley, D. A. (1978c) 'Environmental degradation of the miticide cycloprate (hexadecyl cyclopropanecarboxylate). 4. Beagle dog metabolism', *J. Agric. Food Chem.*, **26**, 76–80.

Quistad, G. B., Staiger, L. E., and Schooley, D. A. (1978b) 'Environmental degradation of the miticide cycloprate. 2. Metabolism by apples and oranges', *J. Agric. Food Chem.*, **26**, 66–70.

Tulliez, J. E., and Bories, G. F. (1979) 'Metabolism of naphthenic hydrocarbons. Utilization of a monocyclic paraffin, dodecylcyclohexane, by rat', *Lipids*, **14**, 292–297.

Yisak, W., Agurell, S., Lindgren, J.-E., and Widman, M. (1978) '*In vivo* metabolites of cannabinol identified as fatty acid conjugates', *J. Pharm. Pharmacol.*, **30**, 462–463.

Progress in Pesticide Biochemistry, Volume 2
Edited by D. H. Hutson and T. R. Roberts
© 1982 John Wiley & Sons, Ltd

CHAPTER 6

The effect of conjugation on the biological activity of foreign compounds in animals

G. D. Paulson

INTRODUCTION

The fact that foreign compounds (drugs, pesticides, antibiotics, and a variety of individual chemicals) are conjugated by animals has been recognized for many years (Williams, 1947, 1959; Smith and Williams, 1966; Young, 1977). Conjugation of functional groups such as $-OH$, $-COOH$, $-SH$, $-NH_2$, and $-NHOH$ with a variety of moieties (acetyl, methyl, glucuronic acid, sulphate, amino acids, glutathione, etc.), is now commonly referred to as type II or phase II metabolism, as defined by Williams (1947).

Early studies in the mid to late 1800s showed that certain conjugated compounds were much less toxic than the parent compound. These and other observations such as the tendency for many conjugates to be rapidly eliminated in the urine led to the general conclusion that conjugation is always a detoxification mechanism. However, more recent research (primarily during the past ten to 15 years) demonstrated that certain compounds are converted to more biologically active compounds by conjugation. These observations have led to a growing awareness that conjugates are not always innocuous compounds. For instance, Caldwell (1978) stated '. . . conjugation reactions are deserving of much more attention from the biochemical pharmacologist than they have received in the past'.

Numerous reviews dealing with the biosynthesis, physical properties, isolation and characterization of conjugates are available (Axelrod, 1962, 1971; Boyland, 1971; Boyland and Chasseaud, 1969; Chasseaud, 1973, 1976a,b; Dodgson and Rose, 1976; Dutton, 1966, 1971; Dutton and Burchell, 1977; Dutton et al., 1977; Hollingworth, 1977; Hutson, 1970, 1972, 1975; Miettinen and Leskinen, 1970; Parke and Smith, 1977; Paulson, 1976, 1980; Weber, 1971; Wood, 1970). This review will be an attempt to summarize some of the more important biological consequences of conjugation of foreign compounds in animals.

THE EFFECT OF CONJUGATION AS EVALUATED BY ACUTE TOXICITY STUDIES

Perhaps the most direct way of measuring the effect of conjugation on the biological activity of a compound is to compare the acute toxicity of the parent compound with the acute toxicity of the conjugate. The author acknowledges that some (perhaps much) pertinent information was unintentionally overlooked; however, the number of publications describing direct comparisons of the acute toxicity of compounds before and after conjugation is quite limited. This is surprising in light of the common, almost casual use of the word 'detoxification' to describe conjugations. A few studies of this nature that have been

Table 1 The effect of conjugation on the acute toxicity of foreign compounds

Compound	Toxicity $(LD_{50})^*$	Conjugate	Toxicity (LD_{50})	Reference
Cyclohexylamine	100 mg kg^{-1}	glucuronic acid	600 mg kg^{-1}	Smith and Williams (1966)
Benzoic acid	2 g kg^{-1}	glycine	4.2 g kg^{-1}	Williams (1959)
p-Aminobenzoic acid	2.8 g kg^{-1}	glycine	4.9 g kg^{-1}	Williams (1959)
Phenylacetic acid	1.5 g kg^{-1}	glycine	4.5 g kg^{-1}	James et al. (1972)
p-Chlorophenyl-acetic acid	0.8 g kg^{-1}	glycine	2.5 g kg^{-1}	James et al. (1972)
Sulphadiazine	1.6 g kg^{-1}	N^4-acetyl	0.6 g kg^{-1}	Williams (1959)
Sulphamethazine	0.9 g kg^{-1}	N^4-acetyl	1.3 g kg^{-1}	Williams (1959)
Pyridine	1.2 g kg^{-1}	N-methyl	0.22 g kg^{-1}	Williams (1959)
Thiophenol	65 mg kg^{-1}	S-methyl	891 mg kg^{-1}	McBain and Menn (1969)

* All data are expressed as the LD_{50} for mice except for thiophenol and its conjugates which are expressed as the LD_{50} for the rat

reported are summarized in Table 1. Seven out of the nine conjugated compounds shown in Table 1 were less toxic than the parent compound and two compounds were made more toxic by conjugation. Therefore, if the limited number of examples shown in Table 1 are representative of all conjugation reactions, these data support the conclusion that most conjugates are less acutely toxic than the parent compounds. However, it also demonstrates that exceptions do occur. For example, N^4-acetylsulphadiazine was shown to be approximately three times as toxic as sulphadiazine.

THE EFFECT OF CONJUGATION ON URINARY EXCRETION

Conjugation reactions have commonly been referred to as 'detoxification' mechanisms, and 'detoxification' has often been used as a descriptive term to describe 'elimination' in the urine (see reviews cited in introduction). The two terms are not synonymous; however, the effect of detoxification (by metabolism of a compound to a less toxic compound) may be indistinguishable from the effect of rapid elimination of a toxic compound (with or without metabolism) as evaluated by the general well being of the animal. Therefore, a brief review of the more important aspects of kidney function as they relate to urinary excretion of foreign compounds is appropriate. For the purposes of this discussion, the nephron (the basic subunit of the kidney) can be considered to be composed of three parts: the glomerulus, the tubule, and the vascular system (afferent arteriole supplying blood to the glomerulus and peritubular capillaries, which drain into the arcuate vein) (Guyton, 1976).

A substantial portion (19% in man) of the blood delivered by the afferent arteriole filters through the glomerulus. The glomerular filtrate (nearly identical to blood plasma except that it is essentially devoid of all substances with a moleculuar weight equal to or greater than plasma proteins) passes into the tubule. In addition, certain compounds are secreted into the tubule (i.e., secreted from the blood directly into the tubules against a concentration gradient). Thus, the fluid in the tubules contains both filtered and actively secreted substances. However, a compound that enters the tubule by either mechanism is not necessarily excreted in the urine. The renal tubular epithelium may be considered to be essentially a lipid membrane, which is permeable to non-polar compounds and impermeable to polar compounds (Peters, 1961). As the glomerular filtrate passes down the tubule, the volume is drastically reduced (more than 99% in man); non-polar compounds passively diffuse from the tubule into the blood via the peritubular capillaries and are, therefore, generally poorly excreted in the urine. In addition, certain compounds (e.g., glucose, amino acids, vitamins, and certain electrolytes) are reabsorbed from the tubules by active transport.

In contrast, when most polar compounds enter the tubules, there is little or no tendency for them to be passively reabsorbed (i.e., they cannot penetrate the lipid membrane of the tubule). This is illustrated by the effect of pH on the excretion of weak acids such as phenobarbital and salicylic acid, which are slowly excreted if the urine is acid (the non-ionized compounds passively diffuse from the tubules back into the bloodstream) but are rapidly excreted if the urine is basic (Peters, 1961). Similarly, organic bases such as quinine, quinacrine, and mecamylamine are more rapidly excreted if the urine is acidic than if it is basic (Peters, 1961). Several types of conjugates commonly formed by animals (glucuronic acid, sulphate esters, amino acid, and mercapturic acid) are present as anions at physiological pH, and are therefore generally expected to be filtered through the glomerulus but not be reabsorbed from the tubule by passive diffusion. Thus, conjugation of non-polar foreign compounds with these types of conjugates tends to enhance their excretion in the urine.

Conjugation may also affect urinary excretion of a compound by altering the effective rate of glomerular filtration. When an animal is challenged with a foreign compound, only that portion of the material which occurs in the plasma as the free form is immediately available for glomerular filtration (Peters, 1961). Factors that affect the rate of glomerular filtration include diffusion into tissues (both intracellular and extracellular spaces); and binding of the compound to macromolecules in the plasma. Diffusion and binding are usually reversible, but the rate of diffusion of the compound (or its metabolites) and the affinity for binding to macromolecules vary greatly, depending upon the nature of the compound. Studies with sulphonamide derivatives showed that binding occurred at hydrophobic sites on serum albumin and that hydrophobic substitution on these drugs enhanced protein binding (Hsu et al., 1974). On the basis of this information, one may argue that the formation of polar glucuronic acid, sulphate ester, amino acid and mercapturic acid conjugates should generally decrease binding to plasma proteins and thereby enhance glomerular filtration and urinary excretion. Although this is probably generally true, there are exceptions. Norling and Hänninen (1980) found that the sulphate ester and glucuronic acid conjugates of 4-methylumbelliferol, phenolphthalein and 4-nitrophenol bound to rat liver subfractions and to bovine serum albumin.

There is also evidence that many amino acid (Guyton, 1976), glucuronic acid (Sperber, 1948), sulphate ester (Peters, 1961; Sperber, 1948; Powell and Curtis, 1966; Curtis et al., 1974; Hearse et al., 1969a,b,c) and N-acetyl (Bevill and Huber, 1977) conjugates are actively secreted into the tubules. For instance, a much studied compound, p-aminohippuric acid (PAH), readily penetrates the glomerular membrane, and it is not reabsorbed by passive diffusion or by active transport. However, almost all the PAH that remains in the blood (after the glomerular filtrate is formed) is actively transported from the peritubular capillaries into the tubule. Thus, approximately 90% of the total PAH is removed from the blood (approximately 19% by glomerular filtration and 71% by active transport) during one pass through the nephron (Guyton, 1976)

On the basis of the preceding discussion, there can be little doubt that the formation of polar conjugates (glucuronic acid, sulphate ester, amino and mercapturic acid) is of great importance in the elimination of many foreign compounds in the urine of animals. This conclusion is substantiated by a tremendous number of reports showing that these four types of conjugates are rapidly excreted in the urine of animals treated with non-polar foreign compounds (Boyland, 1971; Boyland and Chasseaud, 1969; Dodgson and Rose, 1976; Dutton, 1966, 1971; Dutton and Burchell, 1977; Dutton et al., 1977; Hollingworth, 1977; Hutson, 1970, 1972, 1976; Miettinen and Leskinen, 1970; Parke and Smith, 1977; Paulson, 1976, 1980; Wood, 1970). Nevertheless, there are notable exceptions to this concept. For instance, Walle et al. (1979a,b) reported that the glucuronic acid conjugate of propranolol was not rapidly eliminated in the urine. The concentration of the glucuronic acid conjugate was about four times greater than the concentration of propranolol in the plasma of humans treated with propranolol. These workers suggested that the glucuronic acid conjugate of propranolol (shown to be deconjugated in dogs) may serve as a storage pool for propranolol.

Based on the criteria discussed earlier in this section, one can argue that certain types of conjugation (e.g. acetylation of amines and methylation of phenols) would not be expected to enhance urinary excretion and in fact may tend to decrease urinary excretion of certain compounds. For instance, some N^4-acetylated sulphonamide drugs are more firmly bound to albumin than the parent drugs (Goodman and Gilman, 1970); thus, other factors remaining equal, acetylation would be expected to decrease the glomerular filtration rate and therefore decrease urinary excretion of these compounds. In addition, the physical properties of certain conjugated compounds may in themselves be injurious to the animal. For example, N^4-acetylsulphathiazole is less than one-tenth as soluble as sulphathiazole in water at 37 °C (Williams, 1959). A problem with the use of certain sulphonamide drugs, including sulphathiazole, is the formation of crystalline aggregates (usually the N^4-acetyl conjugate, which is the major metabolite of many sulphonamide drugs) in the kidney, ureters and bladder (Bevill and Huber, 1977). Fortunately, this problem has been minimized by the synthesis of structurally modified sulphonamide drugs to improve the solubility of these drugs and their N^4-acetyl conjugates and at the same time retain the desired antibacterial activity.

THE EFFECT OF CONJUGATION ON BILIARY SECRETION OF FOREIGN COMPOUNDS

In contrast to urinary excretion, the secretion of a compound in the bile may or may not be a method of eliminating that compound from the body. Nevertheless, conjugation of a foreign compound may dramatically change the extent of biliary secretion of the compound; therefore, a brief discussion of the major factors controlling biliary secretion is appropriate. With the exception of the formed elements, almost all components of blood readily penetrate the hepatic sinusoids (perisinusoidal space of Disse). The transfer of a compound from the hepatic sinusoids into the bile involves two processes: (i) uptake from the sinusoid into the parenchymal cell and (ii) transfer from the parenchymal cell into the bile canaliculus. Both processes may be accomplished by passive diffusion, active transport, pinocytosis or some combination of these processes, depending on the nature of the compound (Shanker, 1968; Klaassen, 1977; Forker, 1977). Many of the organic compounds secreted in the bile are formed in the liver (Forker, 1977); thus, the properties and mode of transfer of a compound into the parenchymal cell may be much different from the properties and mode of transfer of the metabolite from the parenchymal cell into the bile canaliculus. However, there is general agreement that compounds which occur in the bile at much higher concentrations than in the plasma are actively transported across one and perhaps both membranes (Klaassen, 1977).

Molecular weight is one of two important factors which determine whether or not a compound will be appreciably secreted in the bile (Hirom et al., 1972a,b, 1974; Abou-El-Makarem et al., 1967). There appears to be a minimum molecular weight requirement (threshold), which varies with different animal species. The threshold for organic anions was estimated to be 325 ± 50 for the rat, 400 ± 50 for the guinea pig and 475 ± 50 for the rabbit (Hirom et al., 1972a). Many foreign compounds (drugs, pesticides, and industrial pollutants) have molecular weights in the range of 100–300. Thus, even if the foreign compound meets the requirements for biliary secretion described below, it would not be expected to be excreted in the bile because the molecular weight is below the threshold level. However, conjugation of such compounds with glucuronic acid, glutathione, sulphate, or amino acids often adds sufficient mass so that the molecular weight of the product exceeds the threshold. Conjugates (glycine, acetyl and sulphate esters) of benzene and related compounds (Abou-El-Makarem et al., 1967) and naphthol (Hearse et al., 1969a) with a molecular weight less than 300 were very poorly secreted in rat bile. However, the glucuronic acid conjugate of thiamphenicol, chloramphenicol and phenolphthalein (molecular weights greater than 500) were actively transported into the bile of rats (Uesugi et al., 1974).

In contrast to urine, there does not appear to be a direct relationship between biliary secretion and polarity as measured by solubility in solvents and in vivo studies (Hirom et al., 1972b; Klaassen, 1977). According to Hirom et al. (1974): 'a certain balance between hydrophilic and hydrophobic properties (amphilicity) is required before a compound can cross liver membranes'. That is, a molecule with one region that is hydrophilic and another region that is hydrophobic is a candidate for biliary secretion. Many of the foreign compounds are conjugated are exposed are very non-polar. Thus, when these compounds are conjugated with glucuronic acid, sulphate, amino acids, and glutathione (in which the conjugating moieties exist as anions at physiological pH), the product often has the appropriate degree of amphilicity for efficient biliary secretion. For instance, Chasseaud (1976a) and others have discussed the fact that many glutathione conjugates have ideal physiochemical properties for preferential secretion in the bile.

As previously mentioned, secretion of a compound in bile may or may not be a method for elimination from the body. Some components in bile enter and remain in the intestinal contents and are directly eliminated in the faeces; for such compounds, the effect of biliary secretion is similar to urinary excretion. However, many other things may happen to compounds secreted in bile once they enter the gut (Klaasen, 1977; Bakke *et al.*, 1981). These include: (i) absorption of the unaltered compound from the gut; (ii) further metabolism by the gut (or by bacteria in the gut) followed by absorption from the gut or elimination in the faeces; (iii) further metabolism of compounds (absorbed as described in i and ii) in other tissues in the body; (iv) retention of compounds (absorbed and/or metabolized as described in i–iii) in the tissues; and (v) elimination of some or all of the compounds described in i–iv in the bile or urine.

There is much evidence that many polar conjugates of foreign compounds (glucuronic acid, sulphate esters, amino acid and glutathione) which enter the small intestine in the bile are extensively metabolized (Klaassen, 1977; Bakke *et al.*, 1981). For example, it has been long recognized that glutathione (GSH) conjugates are metabolized to mercapturic acid conjugates, which are commonly excreted in the urine of animals (Boyland and Chasseaud, 1969; Hutson, 1970, 1972, 1975; Grover, 1977; Boyland, 1971; Wood, 1970). However, there is now substantial evidence that the metabolism of GSH and its catabolites (cysteinyl-glycine, cysteine and mercapturic acid conjugates) by the gut and by microflora in the gut is exceedingly complex, resulting in an array of secondary products; moreover, at least some of these secondary products formed in the gut are rapidly reabsorbed and metabolized still further by other tissues in the animal (Bakke *et al.*, 1981).

INDIRECT EVIDENCE FOR DETOXIFICATION OF FOREIGN COMPOUNDS BY CONJUGATION

The conclusion that conjugation of certain foreign compounds is a detoxification mechanism is supported by many types of indirect evidence. One of these is the toxicological consequences of giving animals large doses of the material which overwhelm one or more conjugating systems. For example, many highly electrophilic foreign compounds, if not in some way inactivated, spontaneously react with nucleophilic proteins and nucleic acids and thereby elicit tissue damage and in some instances carcinogenesis: there is much evidence that reaction with glutathione (a strong nucleophile) is the most important mechanism to protect animals from such compounds (Arias and Jakoby 1976). Thus, any factor which renders the glutathione system less efficient (or if the system is overwhelmed with a large dose) can be expected to render the animal more susceptible to electrophilic compounds. There is abundant evidence that this is true (Mitchell and Jollow, 1974, 1975; Thor *et al.*, 1979; Mitchell *et al.* 1973; Thor and Orrenius, 1980; Vina *et al.*, 1980; Potter *et al.*, 1974; Arias and Jakoby, 1976; Timbrell, 1979).

When acetaminophen is consumed in small amounts by humans, it is metabolized primarily by glucuronic acid and sulphate ester conjugation (these conjugates are rapidly eliminated in the urine), and to a much lesser extent to a reactive intermediate, which is subsequently metabolized primarily to mercapturic acid and related conjugates (Figure 1). Levy (1978) presented evidence that the toxicity of large doses of acetaminophen resulted when the sulphate ester and glucuronic acid conjugation systems were overwhelmed and acetaminophen was then preferentially converted to the reactive intermediate, which spontaneously reacted with tissue macromolecules. This caused necrosis of the liver and in severe cases, fatal toxicity. Levy (1978) contends that the glucuronic acid and sulphate ester conjugation systems are 'capacity limited' (transferase enzymes have low K_m values and are saturated at low substrate concentration) and that the conversion of acetaminophen to the reactive intermediate obeys first order kinetics over a broad substrate range. Therefore, when animals are exposed to larger doses, this reaction becomes more important. This, in turn, results in toxicity. Levy (1978) discussed other examples of 'capacity limited' kinetics of foreign compounds. Characteristics of this condition include: (i) the apparent biological half-life of the compound increases with increasingly larger doses; (ii) the total clearance decreases with dose; (iii) the percentage composition of metabolites changes with dose; and (iv) steady state body levels increase more than proportionately with increasing dosing rate.

There is evidence that certain types of nutritional stress depress conjugation systems, which results in increased sensitivity to foreign compounds. Miettinen and Leskinen (1970) cited an example of how a woman consumed a contraceptive compound for an extended time with no apparent adverse effect. However, when this individual fasted and continued to consume the compound at the same rate, severe liver damage resulted. Subsequent studies demonstrated that fasting decreased the ability to form the glucuronic and sulphate ester conjugates of the drug. It was concluded that decreased ability to form these conjugates was responsible for, or contributed to, the observed toxicity. Wood (1970) reviewed studies in which growth inhibition observed in rats fed bromobenzene could be overcome by feeding higher levels of methionine and cysteine (precursors of glutathione).

The ability (or lack of ability) of animals to conjugate certain foreign compounds is directly related to the biological activity of these compounds (Smith and Williams, 1966; Miettinen and Leskinen, 1970; Caldwell, 1979). The ability to form certain types of conjugates (and susceptibility to certain foreign compounds) is quite variable from species to species, and between animals of the

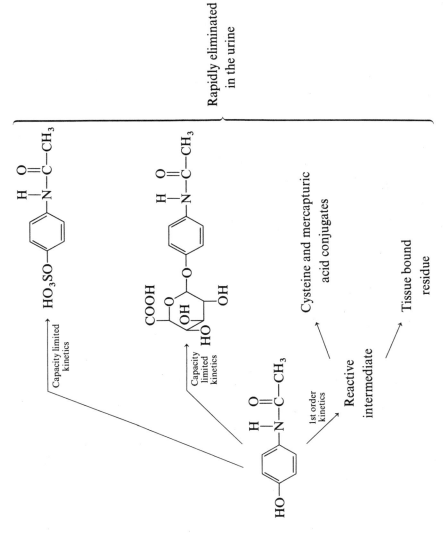

Figure 1   The effect of overloading 'capacity limited' enzyme systems on the toxicity of acetaminophen

same species at different ages or with genetic defects. For instance, Caldwell (1979) indicated that phenol, naphthol, paracetamol and benzoic acid are five to 20 times more toxic to the cat than to the rabbit because the cat has very limited ability to form glucuronic acid conjugates. Furthermore, the sea lamprey has only limited ability to form the glucuronic acid conjugate of 3-trifluoromethyl-4-nitrocatechol, and it is quite susceptible to poisoning by this compound. In contrast, glucuronic acid conjugation is the major route of metabolism of this compound in the trout, which is much less susceptible to poisoning (Lech and Statham, 1975). Other species that have only limited ability to form certain types of conjugates include the pig (sulphate esters), Indian fruit bat (glycine), dog (certain types of N-acetyl conjugates) and guinea pig (mercapturic acids) (Caldwell, 1979). Williams (1974) prepared a comprehensive review of the differences in the conjugation of foreign compounds by man, other primates, and other animals. In general, differences (both qualitative and quantitative) in the conjugation of foreign compounds are the rule rather than the exception.

Many premature and newborn animals have only a very limited ability to form glucuronic acid conjugates (Williams and Millburn, 1975). In one case, more than 30 premature babies died when treated with chloramphenicol. This drug is rapidly conjugated with glucuronic acid in adults. Apparently, the chloramphenicol accumulated in the infants to a lethal concentration because of the infants inability to conjugate and excrete it in the urine.

Animals with limited or no ability to form certain types of conjugates as a result of inborn error are more susceptible to the effects of some foreign compounds than are 'normal animals'. A classic example is the Gunn strain of rats which has only very limited ability to form the glucuronic acid conjugate of phenolic compounds such as menthol and p-aminobenzoic acid (Miettinen and Leskinen, 1970). Animals, including humans, with limited ability to form N-acetyl conjugates ('slow acetylators') are more susceptible to procainamide, hydralazine, phenelzine and salicylazosulphapyridine than are 'fast acetylators' (Drayer and Reidenberg, 1977). Other genetic defects resulting in decreased ability to form conjugates and therefore greater susceptibility to certain foreign compounds include the Crigler–Nazzar syndrome and Gilberts disease (little ability to form glucuronides) and Downs Syndrome (little ability to form glycine conjugates) have been summarized (Caldwell 1979).

THE EFFECT OF CONJUGATION ON THE ACTIVITY OF DRUGS

Most drugs, by design, have at least some degree of complementarity with their site of action. Smith and Williams (1966) suggested that conjugated drugs are often inactive because they cannot penetrate membranes and/or because conjugation destroys the complementarity of the compound with their site of action. There is evidence that conjugation does indeed destroy or diminish the activity

Table 2   The effect of conjugation on the activity of selected drugs

Compound	Acitivity	Conjugate	Activity	Reference
Trichloroethanol	hypnotic	glucuronic acid	very low hypnotic activity	Smith and Williams (1966)
Glutethimide	hypnotic	glucuronic acid	no hypnotic activity	Keberle *et al.* (1962)
Meprobamate	tranquilizer	glucuronic acid	no tranquilizer activity	Smith and Williams (1966)
4-Aminoantipyrine	analgesic	acetyl	no analgesic activity	Williams (1959)
Antipyrine	analgesic	glucuronic acid	no analgesic activity	Smith and Williams (1966)
4-Hydroxy-acetanilide	analgesic	glucuronic acid	no analgesic activity	Williams (1959)
Phlorizin	glycosuric	glucuronic acid	glycosuric	Smith and Williams (1966)
Oestriol glucuronide	oestrogen	glucuronic acid	oestrogen	Smith and Williams (1966)
Diethylstillboestrol	oestrogen	glucuronic acid	weak oestrogen	Smith and Williams (1966)
Hexoestrol	oestrogen	glucuronic acid	weak oestrogen	Smith and Williams (1966)

of many drugs. For example, activity of the first six compounds shown in Table 2 was reduced or destroyed by conjugation. Furthermore, there is evidence that when conjugation systems are inhibited, the activity of certain drugs is prolonged. For instance, Smith and Williams (1966) discussed how inhibitors of glucuronic acid conjugation (novobiocin, SKF 525A and morphine) prolong the activity of myonesin, amphetamine and chloral hydrate.

However, in other cases (see lower part of Table 2), conjugated drugs retained at least part of the biological activity of the parent compound. Thus, it cannot be assumed that conjugation destroys biological activity. It may, in fact, enhance biological activity. For instance, it has been demonstrated that certain conjugates of morphine are more biologically active than morphine, and the type of activity that is altered depends on (i) the site of conjugation and (ii) the type of conjugate (Mori *et al.*, 1972; Shimomura *et al.*, 1971; Labella *et al.*, 1979; Yoshimura *et al.*, 1972). Morphine-6-glucuronide and morphine-6-sulphate have much stronger and longer lasting analgesic activity than morphine. However, morphine-3-glucuronide has less analgesic activity than morphine (Shimomura *et al.*, 1971; Mori *et al.*, 1972; Oguri, 1980).

In contrast to their relative analgesic activities, morphine-3-glucuronide and morphine-3-sulphate are several hundred times more active than morphine in causing hyperactive motor activity (central excitatory activity) and convulsions

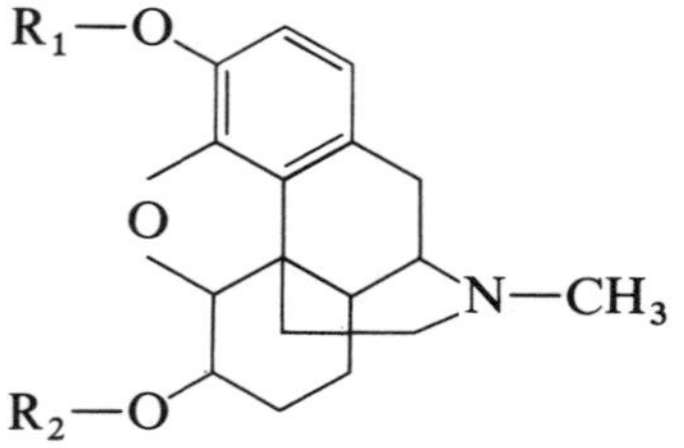

R_1	R_2	
Glucuronic acid	H	high activity
SO_3H	H	high activity
$C(=O)—CH_3$	$C(=O)—CH_3$ (heroin)	high activity
H	$C(=O)—CH_3$	intermediate activity
$C(=O)—CH_3$	H	intermediate activity
H	H (morphine)	lower activity
CH_3	H (codeine)	lower activity
CH_3	CH_3 (thebaine)	lower activity

Figure 2 Excitatory activity of morphine and derivatives in rats

in animals (Figure 2) (Labella *et al.*, 1979). Acetylation of the 3 and 6 position resulted in compounds with intermediate activity, and methylation of the 3 and 6 position gave compounds with excitatory activity similar to morphine.

Eisner and Shahidi (1972) reported an interesting case in which a conjugate elicited an immune response but the parent compound did not. A person consuming *N*-acetyl-*p*-aminophenol (NAPA) developed thrombocytopenia; the immune response, however, was not caused by NAPA but was due to the sulphate ester conjugate. The antigen in the blood did not react with NAPA or the glucuronic acid conjugate of NAPA.

THE EFFECT OF HYDROLYSIS AND OTHER TYPES OF METABOLISM ON THE ACTIVITY OF FOREIGN COMPOUNDS

Some authors have questioned whether conjugates *per se* have biological activity and contend that any activity that they possess results from hydrolysis of the conjugate. For example, Smith and Williams (1966) stated '. . . it is doubtful whether any glucuronide examined possesses significant pharmacological activity or toxicity of its own, for there is always some information available which suggests that if a glucuronide shows activity this may well be due to its conversion to the corresponding aglycone, even though this may only occur to a minor extent'. With the exception of glutathione and related conjugates, there are enzymes or enzyme systems in animal tissues capable of hydrolysing all of the major classes of conjugates. Indeed, the fact that animal cells contain substantial enzymatic activity for both the formation and hydrolysis of conjugates resulted in questioning the conclusion that conjugates are inactive end products of metabolism. Dodgson and Rose (1976) raised the question 'why should a mammalian organism go to great lengths to protect itself by sulphoconjugating phenols and possess at the same time enzymes that are theoretically able to release the phenol from the conjugate?'. Lysosomes and microsomes in a wide variety of animal tissues contain several different hydrolases. However, the direct evidence for this hydrolytic activity has been obtained primarily by *in vitro* studies; therefore, other questions arise. For instance, even though the hydrolytic enzymes can be easily demonstrated *in vitro*, are these enzymes expressed in the intact animal? Are these hydrolytic enzymes compartmentalized in such a manner that once a conjugate is formed, it is not subjected to the hydrolytic activity?

Unfortunately, these and related questions are not answered easily or unambiguously. For instance, if an animal is given a synthetic conjugate orally, this does not necessarily simulate the exposure of a cell (or subcellular component) to the same conjugate that is synthesized at one or more sites within the animal. If it is hydrolysed when given orally, it could be due to bacterial action in the gut and have little or nothing to do with what occurs in the tissues of the animal. On the other hand, even if only the conjugated compound is detected in the peripheral blood and excreta from an orally dosed animal, this does not necessarily prove that some tissue(s) was not exposed to its hydrolysis product. For instance, the conjugate may be hydrolysed in the gut, the hydrolysis product absorbed, and then reconjugated in the liver. In the process, one or more tissues would be exposed to the hydrolysis product. Similarly, if animals are given conjugates by other methods (intravenous, intramuscular, intraperitoneal, etc.), the results may be questioned because they may or may not simulate the normal disposition of conjugates synthesized *in vivo*.

In spite of these limitations, there is considerable evidence that many conjugated foreign compounds are hydrolysed and/or further metabolized in animal systems. The degree to which this occurs is quite variable and depends on the animal, the type of foreign compound, the type of conjugate and probably other factors. For instance, there is apparently little or no deacetylation of N^4-acetylsulphonilamide and N^4-acetylsulphamethazine in rabbits (Gordon et al., 1973). In contrast, the chicken deacetylated 62–83% of administered N^4-acetylsulphonilamide and related sulpha drugs (Shaffer and Bieter, 1950).

Hawkins and Young (1954) gave [^{35}S]phenyl sulphate, 1-naphthyl sulphate, and 2-naphthyl sulphate to rats by intraperitoneal injection and by stomach tube. In all cases, more than 70% of the ^{35}S was excreted as the intact sulphate ester. In contrast, when [^{35}S]-p-nitrophenyl sulphate was given to mice (ip), it was extensively metabolized to [^{35}S]sulphate and several other unidentified [^{35}S]-labelled compounds (Dodgson and Tudball, 1960). When the sulphate esters of 4-hydroxybiphenyl, 4-cyclohexylphenol, 2-cyclohexylphenol and 5,6,7,8-tetrahydro-2-naphthol were given to rats, all were extensively metabolized (Hearse et al., 1969b,c). When [^{35}S]dodecylsulphate was given to rats (ip and oral), metabolites of this compound included [^{35}S]sulphate and butyric acid-4-[^{35}S]sulphate (Denner et al., 1969). Rats converted benzyl sulphate and 1-menaphthyl sulphate to benzylmercapturic acid and 1-menaphthylmercapturic acid (Clapp and Young, 1970). The sulphate ester of dopamine was extensively metabolized by the dog, rat and guinea pig (Merits, 1976).

Walle et al. (1979a,b) reported that the glucuronic acid conjugate of propranolol was partially hydrolysed to propranolol in animal systems. Goldberg et al. (1977) reported that when the glucuronic acid conjugate of iopanoic acid was perfused into the small intestine of dogs, an average of 31 % of the dose was recovered in the bile as the glucuronic acid conjugate. Although their results were not conclusive, there was evidence that the compound was hydrolysed in the gut, absorbed as iopanoic acid and then reconjugated with glucuronic acid.

Boyland and Chasseaud (1969) and Wood (1970) cited several examples of animal metabolism of mercapturic acid conjugates including hydrolysis or exchange of the conjugating moiety. Glutathione conjugates are metabolized to a variety of products (Collucci and Buyske, 1965; Anderson and Schultze 1965; Bakke et al., 1981) and methylated foreign compounds (methoxy compounds in particular) are also further metabolized in animals (Axelrod, 1962).

The consequences of the metabolism of conjugated foreign compounds may be good or bad, as will be illustrated by the following examples. Allen et al. (1957) reported that the glucuronic acid conjugate of 2-amino-1-naphthol produced more tumours than 2-amino-1-naphthol when injected into the bladder of mice; they attributed this to the slow release of 2-amino-1-naphthol by β-glucuronidase activity in the bladder. When Marshall and Dorough (1979) dosed rats with [^{14}C]carbofuran, much of the material was secreted in the bile as the glucuronic acid conjugate of [^{14}C]-3-hydroxycarbofuran; however, the urine was the primary route of excretion of ^{14}C, and only a small amount was excreted in the faeces. They concluded that the glucuronic acid conjugate was hydrolysed (presumably to 3-hydroxycarbofuran which is a cholinesterase inhibitor) and reabsorbed from the gut. Baba et al. (1978) took advantage of conjugate hydrolysis to deliver a chemotherapeutic agent. They found that 5-fluorouracil-O-β-D-glucuronide (FU-O-G) was much less toxic (LD$_{50}$ about 5 g kg^{-1}) than 5-fluorouracil. However, FU-O-G was a very effective inhibitor of a transplanted tumour in mice. The β-glucuronidase activity was higher in the tumour than non-cancerous tissue; thus, they concluded that 5-fluorouracil was selectively liberated from FU-O-G in the cancerous tissue and thereby inhibited the growth of the tumour. Thus, administration of the relatively non-toxic FU-O-G was an effective way of delivering 5-fluorouracil to a tumour without subjecting the entire organism to the toxic effects of the latter compound.

Although there are many yet unanswered questions, it is increasingly apparent that conjugates are not always inert metabolic end products. Therefore, it follows that the biological activity (if any) of a conjugate may, in fact be due to one or more metabolites of that conjugate.

THE ROLE OF CONJUGATION IN CARCINOGENESIS, MUTAGENESIS AND CYTOTOXICITY

Chemicals known to induce cancer in animals are a rather diverse group of compounds, but there is one important generalization about these compounds that can be made. That is, they are either strong electrophilic reactants or they are converted to metabolites that are strong electrophilic reactants, and these electrophiles spontaneously react with nucleophilic proteins and nucleic acids (Miller and Miller, 1976, 1977; Caldwell, 1978). Compounds which are not inherently carcinogenic on the basis of their own properties (precarcinogens) may be converted to reactive electrophilic metabolites (ultimate carcinogens) *in vivo* (Miller and Miller, 1976, 1977). There is now persuasive evidence that conjugation is involved in the formation of several ultimate carcinogens.

It has been known for many years that certain aromatic amines, and compounds that can be converted to aromatic amines, are carcinogenic and mutagenic. As Arcos and Argus (1974) and others have discussed, the most active amines are dicyclic and tricylic aromatic amines with the NH_2 moiety on the terminal carbon atom of the longest conjugated chain. Free aromatic amines and their N-acetyl derivatives often have similar activity because most animal systems establish an equilibrium between the amine and its acetylation product (Irving, 1979). Intensive studies have been conducted with 2-aminofluorene, 4-aminobiphenyl, 2-naphthylamine and other related compounds to determine how these compounds induce cancer in animals including man (Miller and Miller, 1969; Miller, 1970; Irving, 1970, 1971, 1978, 1979; DeBaun *et al.*, 1970; Weisburger *et al.*, 1972; Weisburger and Weisburger, 1973). These studies have shown that the first step in the activation of these compounds is N-hydroxylation, and that certain free arylhydroxylamines are toxic and probably carcinogenic (Irving, 1970; Kiese, 1966). However, there is convincing evidence that several N-hydroxy compounds are further activated, through a variety of conjugation reactions, to produce compounds that spontaneously react with tissue nucleophiles resulting in carcinogenesis (Miller and Miller, 1969; Miller, 1970). Several different types of reactive conjugates of N-arylhydroxylamines and N-acyl-N-arylhydroxylamines have been studied and classified (Irving, 1970, 1978); they include: (i) the O-sulphonate (N-sulphate) conjugates of N-acetyle-N-arylhydroxylamines; (ii) the O-acyl derivatives of N-arylhydroxylamines; (iii) the O-glucuronic acid conjugates of N-arylhydroxylamines; (iv) the O-glucuronic acid conjugates of N-acetyl-N-arylhydroxylamines; and

(v) the *N*-glucuronic acid conjugates of *N*-arylhydroxylamines. Formation of each of these types of conjugates (using the animal metabolism of 2-aminofluorene as an example) and the fate of these conjugates is summarized in Figure 3. Each class of conjugates will be discussed individually.

The activation of 2-aminofluorene by sequential acetylation, oxidation, and sulphate ester conjugation and the subsequent spontaneous reaction of this conjugate with tissue macromolecules have been studied in detail. It should be emphasized that the highly reactive sulphate ester conjugate shown in Figure 3 has not been isolated from an *in vivo* system. DeBaun *et al.* (1967) estimated that the half-life of this compound in water is less than one minute. Nevertheless, the evidence obtained by a series of ingenious *in vitro* and *in vivo* studies clearly indicates that the reactive sulphate ester is formed *in vivo* (Miller and Miller, 1969; Miller, 1970; Irving, 1970, 1971, 1978, 1979; DeBaun *et al.*, 1970; Weisburger *et al.*, 1972; Weisburger and Weisburger, 1973) and that it reacts spontaneously with protein, RNA and DNA, leading to the attachment of 2-acetylaminofluorene and 2-aminofluorene residues to these macromolecules. Methionine, cysteine, tryptophan, tyrosine and guanine are the primary sites of attachment for these residues in proteins and nucleic acid (Irving, 1971). *N*-Hydroxy-*N*-acetyl-2-aminofluorene sulphotransferase, the enzyme(s) responsible for the sulphate ester conjugation of *N*-hydroxy-*N*-acetyl-2-aminofluorene, has been partially purified and requires PAPS. The level of *N*-hydroxy-*N*-acetyl-2-aminofluorene sulphotrasferase in the liver correlates well with the LD_{50}, the relative hepatotoxicity, and hepatocarcinogenecity of *N*-hydroxy-*N*-acetyl-2-aminofluorene in experimental animals (Irving, 1975; DeBaun *et al.*, 1970). There is evidence that several other *N*-hydroxyarylamines, including *N*-hydroxy-*N*-methyl-4-aminoazobenzene (Kadlubar *et al.*, 1976), *N*-hydroxyphenacetin, *N*-hydroxy-*N*-acetyl-2-aminonaphthalene (Mulder *et al.*, 1977), *N*-hydroxy-*N*-acetyl-4-aminobiphenyl, *N*-hydroxy-*N*-acetyl-4-aminostilbene, *N*-hydroxy-*N*-acetyl-4-aminoazobenzene and *N*-hydroxy-*N*-acetyl-2-aminophenathrene (Irving, 1978) are also activated by sulphate ester conjugation.

The *O*-glucuronic acid conjugates of *N*-acetyl-arylhydroxylamines which have been studied include the *O*-glucuronic acid conjugate of *N*-hydroxy-*N*-acetyl-2-aminofluorene, *N*-hydroxy-*N*-acetyl-4-aminostillbene, *N*-hydroxy-*N*-acetyl-4-aminobiphenyl, *N*-hydroxy-*N*-acetyl-2-aminophenanthrene, *N*-acetyl-*N*-phenyl-hydroxylamine, *N*-hydroxy-*N*-acetyl-2-aminonaphthalene (Irving, 1977, 1978; Irving and Wiseman, 1971) and *N*-hydroxyphenacetin (Hinson *et al.*, 1979). In contrast to the sulphate ester conjugates, the *O*-glucuronic acid conjugates of *N*-acetyl arylhydroxylamines are sufficiently stable to be excreted in the urine and secreted in the bile. For instance, about 30 % of the *N*-acetyl-2-aminofluorene given orally to rabbits was eliminated in the urine within 24 h as the *O*-glucuronic acid conjugate of *N*-hydroxy-*N*-acetyl-2-aminofluorene (Irving, 1978). However, certain compounds of this class or their metabolites

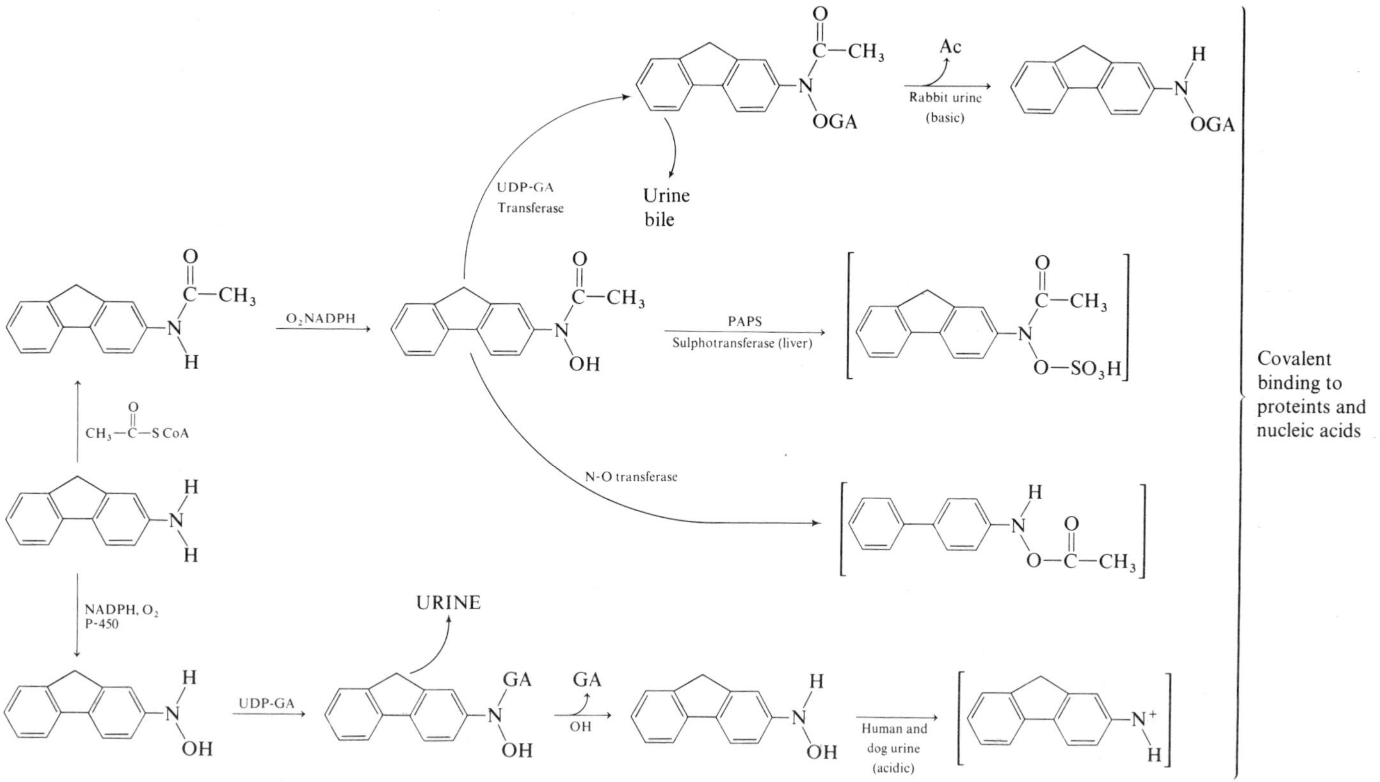

Figure 3 The effect of conjugation of aminofluorene on binding to nucleophilic macromolecules. (GA = glucuronic acid)

also react spontaneously with tissue macromolecules, and the reaction products are similar to those observed from the reaction of the sulphate ester of N-acetyl-arylhydroxylamines, except for the loss of the acetyl moiety of the reacting O-glucuronide (Irving, 1970, 1978). The nature of the aryl moiety has a marked effect on the reactivity of this class of conjugates with tissue macromolecules. Irving (1977) reported that the relative reactivity (highest to lowest) of the O-glucuronic acid conjugates of several N-acetyl-arylhydroxylamines was as follows: N-hydroxy-N-acetyl-2-aminofluorene, N-hydroxy-N-acetyl-4-amino-stillbene, N-hydroxy-N-acetyl-4-aminobiphenyl and N-hydroxy-N-acetyl-2-aminophenanthrene. The reactivity of these compounds is elevated at slightly basic pH. This may explain why the urinary tract of the rabbits (pH 8.5–9.0) is especially susceptible to N-acetyl-2-aminofluorene and N-hydroxy-N-acetyl-2-aminofluorine carcinogenesis. Irving (1979) proposed that these compounds are metabolized to the O-glucuronic acid conjugate of N-hydroxy-N-acetyl-2-aminofluorene and that this conjugate reacts spontaneously with the epithelium of rabbit urinary tract.

There is evidence that the glucuronic acid conjugates of N-acetylarylhydroxylamines may be further activated by removal of the N-acetyl group (Figure 3). The O-glucuronic acid conjugate of N-hydroxy-N-acetyl-2-aminofluorene was deacetylated to give a more reactive product (i.e. the O-glucuronic acid conjugate of N-hydroxy-2-aminofluorene) (Cardona and King, 1976). The latter compound was mutagenic and reacted spontaneously with nucleic acids, methionine, and tryptophan (Irving and Russel, 1970; Irving, 1978). Mulder et al. (1977) reported that the O-glucuronide of N-hydroxyphenacetin reacted with proteins to form covalently bound residues and suggested that this may be an explanation for the toxicity of high doses of phenacetin to the kidney and bladder.

Another class of related compounds is the N-glucuronic acid conjugates of N-arylhydroxylamines (Figure 3); those that have been investigated include the N-glucuronic acid conjugate of N-1-naphthylhydroxylamine, N-2-napthyl-hydroxylamine, N-4-biphenylhydroxylamine, N-hydroxy-2-aminofluorene, N-hydroxy-4-aminoazobenzene (Kadlubar et al., 1977, 1978) and N-hydroxy-4-aminobiphenyl (Radomski et al., 1977). These compounds are relatively stable and non-reactive at neutral pH. Under acid conditions (pH 5), however, the N-glucuronic acid conjugates of N-arylhydroxylamines decompose to free arylhydroxylamines which may be activated to compounds capable of binding to tissue macromolecules. Kadlubar et al. (1977, 1978) proposed that N-glucuronic acid conjugates of this type may be 'transported' to the bladder, hydrolyzed in the acidic urine of dogs and humans, and the free arylhydroxylamines converted to arylnitrenium ions (Figure 3). They proposed that the latter compounds then spontaneously react with tissue macromolecules leading to cancer of the bladder.

Certain arylhydroxylamine-O-acetyl conjugates react spontaneously with tissue macromolecules and are carcinogenic (Bartsch et al., 1972, 1973; King, 1974; King et al., 1976; King and Allaben, 1978; King and Olive, 1975; King et al., 1978; Lotlikar, 1972). Compounds of this type have not been isolated intact because of their highly reactive nature. However, evidence for their formation has been obtained by 'trapping studies' with nucleophiles such as methionine, guanosine and nucleic acids. Arylhydroxylamine-O-acetyl conjugates are biosynthesized from N-acetyl-arylhydroxylamines under the influence of an N-O-acyltransferase (i.e. transfer of the N-acetyl moiety in N-acetylarylhydroxylamine to form the more reactive O-acyl derivative of the N-arylhydroxylamine as illustrated in Figure 3). The N-O-acyltransferase responsible for this activation has been detected in several tissues that are susceptible to arylamine-induced carcinogenesis (King and Allaben, 1978). There is evidence for the activation of several compounds by the N-O-acyltransferase mechanism. These include the N-hydroxy derivatives of N-acetyl-2-aminofluorene, N-acetyl-4-aminobiphenyl, diacetylbenzidine, N-acetyl-4-aminostillbene, N-acetyl-2-aminophenanthrene and N-acetyl-2-aminonapthalene (King and Alaben, 1978).

The mechanisms involved in the binding of activated arylamines to tissue macromolecules may vary, depending on the aryl moiety involved. Hinson et al. (1979) proposed that the first step is cleavage of the N–O bond to form the nitrenium ion. In some cases (e.g. nitrenium ion of acetylaminofluorene), the nitrenium ion is the reactive compound that reacts with nucleophilic proteins and nucleic acids. In other cases (e.g. the nitrenium ion of phenacetin), the nitrenium ion does not react directly with tissue macromolecules; it is first converted to the N-acetylimidoquinone, and the latter compound reacts with tissue proteins and nucleic acids (Hinson et al., 1978). Several different mechanisms may be involved in the activation of an aromatic amine within an animal species. In addition, animals of different species may metabolize an aromatic amine differently, depending on their enzymatic capabilities (ability or lack of ability to N-acetylate, N-O acetyltransferase activity, etc.). Other factors peculiar to a certain species (pH of urine, etc.) may also have profound effects on the toxicity, mutagenicity and carcinogenicity of these metabolites. Adding to the complexity are the observations that the physiological condition of the animal and pretreatment of the animal may also effect the metabolism and biological activity of the aromatic amines (Irving, 1970, 1971, 1978, 1979; Miller 1970; Miller and Miller 1969; DeBaun et al., 1970; Weisburger et al., 1972, Weisburger and Weisburger, 1973).

Certain methylated polycyclic aromatic hydrocarbons are stronger carcinogens than the corresponding unsubstituted compounds, and when the methyl group is oxidized to a hydroxymethyl (a conversion that occurs in animals), the products are even stronger carcinogens. For instance, Cavalieri et al. (1978) reported that 7-hydroxymethylbenz[a]anthracene is a stronger carcinogen than 7-methylbenz[a]anthracene. However, there is now evidence that the hydroxymethyl compounds are further activated by conjugation. For instance, the sulphate ester of 6-hydroxymethylbenzo[a]pyrene is a stronger carcinogen than 6-hydroxymethylbenzo[a]pyrene. This observation led to the proposal that hydroxymethylpolycyclic hydrocarbons are converted to carcinogens by esterification in animal systems. A recent report by Brauns et al. (1980) supports this proposal. They found that salicylamide (an inhibitor of sulphate ester conjugation) prevented the activation of 7,12-dimethylbenz[a]-anthracene as measured by the DNA excision repair in isolated rat hepatocytes. It was proposed that 7-methylbenz[a]anthracene is oxidized to hydroxy-methylbenz[a]anthracene, and the latter compound is conjugated with sulphate to form the 'ultimate electrophilic metabolite' responsible for the carcinogenic activity. In light of these results, another report by Jenner and Testa (1978) may be especially important. They found that benzo[a]pyrene was converted to 6-hydroxymethylbenzo[a]pyrene by rat liver microsomes in a reaction that apparently did not involve 6-methylbenzo[a]pyrene as an intermediate. It was suggested that this mechanism may be involved in the carcinogenic activity of certain polycylic hydrocarbons.

Safrole, a weak hepatocarcinogen, is oxidized in animal systems to 1'-hydroxysafrole, which is more carcinogenic then the parent compound (Wislocki et al., 1976). Although the intermediate has not been isolated, there is evidence that 1'-hydroxysafrole is converted to the sulphate ester (Figure 4) and that this compound spontaneously reacts with tissue nucleophiles. When 1'-hydroxysafrole was incubated with rat and mouse liver cytosols and PAPS, the compound was activated and bound to RNA (Wislocki et al., 1976). Synthetic 1'-acetoxysafrole is a reactive electrophile (reacts with nucleosides and methi-

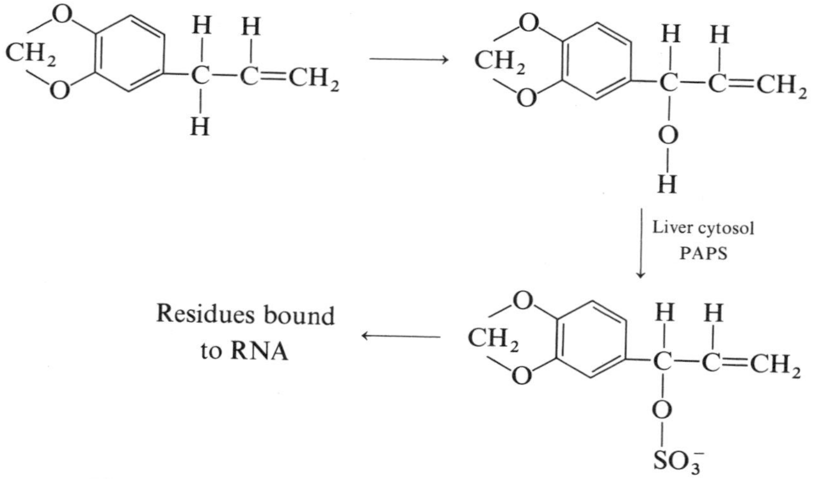

Figure 4 Activation of safrole by oxidation and conjugation

onine) and is carcinogenic; however, there is apparently no evidence for the formation of this compound in animal systems.

Kinoshita and Gelboin (1978) suggested that glucuronic acid conjugates of hydroxylated benzo[a]pyrene may be transported to tissues distal to their formation and then converted to carcinogens. They reported that during the hydrolysis of benzo[a]pyrene-3-glucuronide by β-glucuronidase, a reactive intermediate was formed that covalently bound with DNA. The binding was not due to free 3-hydroxy-benzo[a]pyrene which is not carcinogenic and only weakly mutagenic. The binding of 3-hydroxybenzo[a]pyrene to DNA was only one-tenth that observed when benzo[a]pyrene-3-glucuronide was incubated with β-glucuronidase.

Isoniazid, a compound that has been widely and successfully used to treat tuberculosis, is hepatotoxic to some individuals (Mitchell *et al.*, 1976a,b; Timbrell, 1979). Studies by Mitchell *et al.* (1976a,b) indicated a relationship between isoniazid hepatotoxicity and the rate of acetylation of this compound. The metabolism of isoniazid (Figure 5) in susceptible individuals was characterized by rapid formation of acetylisoniazid, which was hydrolysed to yield isonicotinic acid and acetylhydrazine. Isoniazid, acetylisoniazid, acetylhydrazine and hydrazine were tested for hepatotoxicity in rats; only acetylisoniazid and acetylhydrazine caused cell necrosis. Because of this and other supporting information, they concluded that acetylation is the crucial step in the activation of isoniazid to a hepatotoxic metabolite(s). This conclusion is supported by epidemiology studies which showed that individuals with greater ability to form *N*-acetyl conjugates ('fast acetylators') were more susceptible than 'slow acetylators' to isoniazid hepatotoxicity (Mitchell *et al.*, 1976a,b). It is interesting that 48% of white Americans have 'fast acetylator' phenotypes, 88% of the Japanese have this phenotype· and that orientals are more susceptible to isoniazid toxicity.

As discussed in a previous section, there is abundant evidence that glutathione conjugation is of crucial importance in the detoxification of many foreign

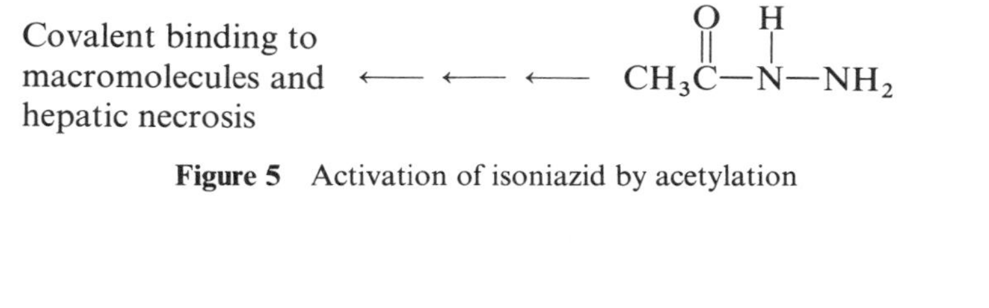

Figure 5 Activation of isoniazid by acetylation

compounds. However, there is now evidence that glutathione conjugation of certain compounds may be detrimental. Studies by van Bladeren *et al.* (1979) indicated that certain disubstituted cyclohexanes were converted to powerful mutagens by glutathione conjugation. They concluded that 'in principal any compound substituted with good leaving groups on vicinal carbon atoms can be converted into a potential mutagen by glutathione conjugation'.

CONCLUDING REMARKS

As the pool of information about conjugates of foreign compounds continues to grow, it is obvious that broad all-inclusive statements about the biological consequences of conjugation of foreign compounds in animals are not appropriate. The activity of these compounds is quite variable depending on the nature of the compound, the conjugating moieties, and other factors. Therefore, the use of 'detoxification' and 'deactivation' as inclusive terms to describe all conjugation reactions is not justified and is in fact misleading. That is not to say that conjugation is not of importance in the detoxification of many foreign compounds. Although it cannot be classified as detoxification *per se*, the greater tendency of foreign compounds to be excreted in the urine after conjugation is also an extremely important factor. In fact, there is general agreement that both deactivation and enhanced urinary excretion as consequences of conjugation are of crucial importance in protecting animals from the adverse effects of many foreign compounds.

One point is clear, however, not all conjugations of foreign compounds result in less active products. Some conjugates are equally active and some are much more active than the parent compound. The conversion of 'precarcinogens' to 'ultimate carcinogens' is the most dramatic example of activation by conjugation, but there are other examples as well.

It is true that many conjugated foreign compounds are rapidly excreted in the urine with little or no further metabolism once they are formed. In these cases, conjugates can be considered as metabolic end products. However, many conjugates may be further metabolized in a variety of ways. These include: (i) hydrolysis of the conjugate to regenerate the parent compound; (ii) hydrolysis followed by reconjugation with the same or a different moiety; and (iii) further metabolism of the intact conjugated compound. If a conjugate is secreted in the bile, it may be reabsorbed from the gut or eliminated in the faeces without modification; however, it may also be further metabolized in the gut or by microorganisms in the gut. Finally, metabolites formed in the gut may be absorbed and, in some cases, further metabolized in the animal tissues. Thus, the activity of a conjugated compound (as evaluated by feeding studies, etc.) may, in fact, be due to one or more metabolites of the conjugate.

Finally, it is apparent that there are many unanswered questions regarding the biological significance of the conjugation of foreign compounds. In addition to the types of conjugates that have been recognized for many years, reports of new and unique conjugates of foreign compounds in animal systems continue to appear in the literature. New conjugations reported in the literature in the mid to late 1970s and summarized by Jenner and Testa (1978) include: methylation of aliphatic amines, alcohols and monomethylated catechols; hydroxymethylation of benzo[a]pyrene, benzene and aniline; sulphamation of dapsone in the presence and absence of acetylation; the formation of N-glucuronides of aliphatic amines, quanternary ammonium N-glucuronides, C-glucuronides ($C_{1'}$ of the glucuronide directly attached to carbon of foreign compound), N-glucosides and N-ribosides; methylation, formylation and acetylation of a secondary amine; taurine conjugation of foreign compounds; methylthio conjugation (derived from either GSH or mercapturic acid conjugates or by direct incorporation of methylthio group); phosphate conjugation of 2-naphthylamine and phenol; and fatty acid conjugation. Recently, Crayford and Hutson (1980a) reported that a small percentage of 3-phenoxybenzoic acid was converted to two unique neutral conjugates [2- and 3-(phenoxybenzoyl)-dipalmitins]. To the author's knowledge, these new classes of conjugates formed in animals have not been evaluated for biological activity. Finally, animals are exposed to conjugates (glucose, amino acid, glutathione and related compounds) formed in plants. A limited number of studies have been conducted to determine the metabolic fate of these compounds and their toxicological properties (Dorough, 1976, 1979, 1980; Crayford and Hutson, 1980b). However, many questions about the biological activity of these compounds in animal systems remain unanswered.

REFERENCES

Abou-El-Makarem, M. M., Millburn, P., Smith, R. L., and Williams, R. T. (1967) 'Biliary excretion of foreign compounds', *Biochem. J.*, **105**, 1269–1274.

Arcos, J. C., and Argus, M. F. (1974) *Chemical Induction of Cancer. Structural Basis and Biological Mechanisms*, Academic Press, New York, Vol. IIB.

Allen, M. J., Boyland, E., Dukes, C. E., Horning, E. S., and Watson, J. G. (1957) 'Cancer of the urinary bladder induced in mice with metabolites of aromatic amines and tryptophane', *Brit. J. Cancer*, **11**, 212–228.

Anderson, P. M., and Schultze, M. O. (1965) 'Cleavage of S-(1,2-dichlorovinyl)-L-cysteine by an enzyme of bovine origin', *Arch. Biochem. Biophys.*, **111**, 593–602.

Arias, M., and Jakoby, W. B. (1976) *Glutathione: Metabolism and Function*, Raven Press, New York.

Axelrod, J. (1962). 'Demethylation and methylation of drugs and physiologically active compounds', in *Metabolic Factors Controlling Duration of Drug Action* (B. B. Brodie and E. G. Erdos, Eds.), Pergamon Press, New York, Vol. 6, pp. 97–107.

Axelrod, J. (1971) 'Methyl transferase enzymes in the metabolism of physiologically active compounds and drugs', in *Concepts in Biochemical Pharmacology* (B. B. Brodie and J. R. Gillette, Eds.), Springer-Verlag, New York, Part II, pp. 609–619.

Baba, T., Kidera, Y., Kimura, N. T., Aoki, K., Kamura, T., Taniguchi, S., and Nishikawa, K. (1978) '5-Fluorouracil O-β-D-glucuronide as a newly synthesized chemically modified, nontoxic anticancer drug', *Gann*, **69**, 283–284.

Bakke, J. E., Larsen, G. L., and Aschbacher, P. W. (1981) 'Role of gut microflora in metabolism of glutathione conjugates of xenobiotics', in *Sulfur Chemistry and Biochemistry* (J. Rosen and J. Casida, Eds.), ACS Symposium Series No. 158, pp. 165–178.

Bartsch, H., Dworkin, M., Miller, J. A., and Miller, E. C. (1972) 'Electrophilic N-acetoxy-aminoarenes derived from carcinogenic N-hydroxy-N-acetylaminoarenes by enzymatic deacetylation and transacetylation in liver', *Biochim. Biophys. Acta*, **286**, 272–298.

Bartsch, H., Dworkin, C., Miller, E. C., and Miller, J. A. (1973) 'Formation of electrophilic N-acetoxyarylamines in cytosols from rat mammary gland and other tissues by transacetylation from the carcinogen N-hydroxy-4-acetyl-aminobiphenyl', *Biochim. Biophys. Acta*, **304**, 42–55.

Bevill, R. F. and Huber, W. G. (1977) 'Sulfonamides', *Vet. Pharmacol. Therap.*, **4**, 894–911.

Blunck, J. M., and Crowther, C. E. (1975) 'Enhancement of azo dye carcinogenesis by dietary sodium sulphate', *Eur. J. Cancer*, **11**, 23–31.

Boyland, E. (1971) 'Mercapturic acid conjugation', in *Concepts in Biochemical Pharmacology* (B. B. Brodie and J. R. Gillette, Eds.), Springer-Verlag, New York, Part II, pp. 584–608.

Boyland, E., and Chasseaud, L. F. (1969) 'The role of glutathione and glutathione S-transferases in mercapturic acid biosynthesis', *Adv. Enzymol.*, **32**, 172–219.

Brauns, R. M. E., Van Doorn, R., Bos, R. P., and Henderson, P. T. (1980) 'Inhibition by salicylamide of 7,12-dimethylbenz[a]anthracene—evoked DNA excision repair in isolated rat hepatocytes. An indication of the involvement of a sulfate ester intermediate in carcinogenesis', *Mutation Res.*, **71**, 155–159.

Caldwell, J. (1978) 'The conjugation reactions: The poor relations of drug metabolism', in *Conjugation Reactions in Drug Biotransformation* (A. Aitio, Ed.), Elsevier/North Holland Biomedical Press, Amsterdam, pp. 477–485.

Caldwell, J. (1979) 'The significance of Phase II (conjugation) reactions in drug disposition and toxicity', *Life Sciences*, **24**, 571–578.

Cardona, R., and King, C. M. (1976) 'Activation of the O-glucuronide of the carcinogen N-hydroxy-N-2-fluorenylacetamide by enzymatic deacetylation *in vitro*: formation of fluorenyl-amine-t RNA adducts', *Biochem. Pharmacol.*, **25**, 1051–1056.

Cavalieri, E., Roth, R., Rogan, E., Grandjean, C., and Althoff, J. (1978) 'Mechanisms of tumor initiation by polycyclic aromatic hydrocarbons', in *Carcinogenesis*, (P. W. Jones and R. I. Frendenthal, Eds.), Raven Press, New York, Vol. 3, pp. 273–284.

Chassaud, L. F. (1973) 'The nature and distribution of enzymes catalyzing the conjugation of glutathione with foreign compounds', *Drug Metab. Rev.*, **2**, 185–220.

Chasseaud, L. F. (1976a) 'Conjugation with glutathione and mercapturic acid excretion', in *Glutathione: Metabolism and Function* (M. Arias and W. B. Jakoby, Eds.), Raven Press, New York, pp. 77–114.

Chasseaud, L. F. (1976b) 'Properties of the glutathione S-alkene transferase system catalyzing the conjugation of glutathione with diethyl maleate', in *Glutathione: Metabolism and Function* (M. Arias and W. B. Jakoby, Eds.), Raven Press, New York, pp. 281–284.

Clapp, J. J., and Young, L. (1970) 'Formation of mercapturic acids in rats after the administration of aralkyl esters', *Biochem. J.*, **118**, 765–771.

Collucci, D. F., and Buyske, D. A. (1965) 'The biotransformation of a sulfonamide to a mercaptan and to mercapturic acid and glucuronide conjugates', *Biochem. Pharmacol.*, **14**, 457–466.

Crayford, J. V., and Hutson, D. H. (1980a) 'Xenobiotic triglyceride formation', *Xenobiotica*, **10**, 349–354.

Crayford, J. V., and Hutson, D. H. (1980b) 'The metabolism of 3-phenoxybenzoic acid and its glucoside conjugate in rats', *Xenobiotica*, **10**, 355–364.

Curtis, C. G., Hearse, D. J., and Powell, G. M. (1974) 'Renal excretion of some aryl sulphate esters in the rat', *Xenobiotica*, **4**, 595–600.

DeBaun, J. R., Miller, E. C., and Miller, J. A. (1970) '*N*-hydroxy-2-acetylaminofluorene sulfotransferase: Its probable role in carcinogenesis and in protein (methion-s-yl)-binding in rat liver', *Cancer Res.*, **30**, 577–595.

DeBaun, J. R., Rowley, J. Y., Miller, E. C., and Miller, J. A. (1967) 'Sulfotransferase activation of *N*-hydroxy-2-acetyl-aminofluorene in the rat', *Proc. Soc. Exptl. Biol. Med.*, **129**, 268–273.

Denner, W. H. B., Olavesen, A. H., Powell, G. M., and Dodgson, K. S. (1969) 'The metabolism of potassium dodecyl[^{35}S]sulphate in the rat', *Biochem. J.*, **111**, 43–51.

Dodgson, K. S., and Rose, F. A. (1976) 'Sulfoconjugation and sulfohydrolysis', in *Metabolism Conjugation and Metabolic Hydrolysis* (E. H. Fishman, Eds.), Academic Press, New York, Vol. 1, pp. 239–325.

Dodgson, K. S., and Tudball, N. (1960) 'The metabolic fate of the ester suphate group of potassium *p*-nitrophenyl[^{35}S]sulphate', *Biochem. J.*, **74**, 154–159.

Dorough, H. W. (1976) 'Biological activity of pesticide conjugates', in *Bound and Conjugated Pesticide Residues* (D. D. Kaufman, G. G. Still, G. D. Paulson, and S. K. Bandal, Eds.), Amer. Chem. Soc. Symposium Series No. 29, Washington, D.C., pp. 11–34.

Dorough, H. W. (1979) 'Metabolism of insecticides by conjugation mechanisms', *Pharmacol. Ther.*, **4**, 433–471.

Dorough, H. W. (1980) 'Conjugation reactions of pesticides and their metabolites with sugars', in *Advances in Pesticide Science* (H. Geissbuhler, G. T. Brooks, and P. C. Kearney, Eds.), Pergamon Press, Oxford, pp. 526–536.

Drayer, D. E., and Reidenberg, M. M. (1977) 'Clinical consequences of polymorphic acetylation of basic drugs', *Clin. Pharmacol. Ther.*, **22**, 251–258.

Dutton, G. J. (1966) *Glucuronic Acid, Free and Combined*, Academic Press, New York.

Dutton, G. J. (1971) 'Glucuronide forming enzymes', in *Concepts in Biochemical Pharmacology* (B. B. Brodie and J. R. Gillette, Eds.), Springer-Verlag, New York, vol. 28, pp. 378–400.

Dutton, G. J. and Burchell, B. (1977) 'Newer aspects of glucuronidation', in *Progress in Drug Metabolism* (J. W. Bridges and L. F. Chasseaud, Eds.), John Wiley, New York, vol. 2, pp. 1–60.

Dutton, G. J., Wishart, G. J., Leakey, J. E. A., and Goher, M. A. (1977) 'Conjugation of glucuronic acid and other sugars', in *Drug Metabolism—From Microbe to Man* (D. E. Parke and R. L. Smith, Eds.), Taylor and Francis Ltd., London, pp. 71–90.

Eisner, E. V. and Shahidi, N. T. (1972) 'Immune thrombocytopenia due to a drug metabolite', *New Eng'and J. Med.*, **287**, 376–381.

Forker, E. L. (1977) 'Mechanisms of hepatic bile formation', *Ann. Rev. Physiol.*, **39**, 323–347.

Goldberg, H. I., Lin, S. K., Thoeni, R., Moss, A. A., and Brito, A. (1977) 'Recirculation of of iopanoic acid after conjugation in the liver', *Invest. Radiol.*, **12**, 537–541.

Goodman, L. S. and Gilman, A. (1970) *The Pharmacological Basis of Therapeutics*, 4th edn., Macmillan Co., New York.

Gordon, C. R., Shafizadeh, A. G., and Peters, J. H. (1973) 'Polymorphic acetylation of drugs in rabbits', *Xenobiotica*, **3**, 133–150.

Grover, P. L. (1977) 'Conjugates with glutathione', in *Drug Metabolism from Microbe to Man* (D. V. Parke and R. L. Smith, Eds.), Taylor and Francis, Ltd., London, pp. 105–122.

Guyton, A. C. (1976) *Textbook of Medical Physiology*, 5th edn., W. B. Saunders Company, Philadelphia, pp. 936–944.

Hawkins, J. B., and Young, L. (1954) 'Biochemical studies of toxic agents (5) Observations on the fate of ^{35}S-labelled arylsulphuric acids following their administration to the rat', *Biochem. J.*, **56**, 166–170.

Hearse, D. J., Powell, G. M., and Olavesen, A. H. (1969a) 'The mode of excretion and metabolic fate of potassium naphthyl 2-^{35}S-sulphate', *Biochem. Pharmacol.*, **18**, 197–203.

Hearse, D. J., Powell, G. M., Olavesen, A. H., and Dodgson, K. S. (1969b) 'The application of whole-body autoradiography to a study of the distribution, metabolism and mode of excretion of ^{35}S-labelled aryl sulphate esters', *Biochem. Pharmacol.*, **18**, 205–209.

Hearse, D. J., Powell, G. M., Olavesen, A. H., and Dodgson, K. S. (1969c) 'The influence of some physicochemical factors on the biliary excretion of a series of structurally related sulphate esters', *Biochem. Pharmacol.*, **18**, 181–195.

Hinson, J. A., Andrews, L. S., Mulder, G. J., and Gillette, J. R. (1978) 'Decomposition mechanisms of N-O-glucuronide and N-O-sulfate conjugates of phenacetin and other arylamides', in *Conjugation Reactions in Drug Biotransformations* (A. Aitio, Ed.), Elsevier/North Holland Biomedical Press, Amsterdam, pp. 455–466.

Hinson, J. A., Andrews, L. S., and Gillette, J. R. (1979) 'Kinetic evidence for multiple chemically reactive intermediates in the breakdown of phenacetin N-O-glucuronide', *Pharmacology*, **19**, 237–248.

Hirom, P. C., Millburn, P., Smith, R. L., and Williams, R. T. (1972a) 'Species variations in the threshold molecular-weight factor for the biliary excretion of organic anions', *Biochem. J.*, **129**, 1071–1077.

Hirom, P. C., Millburn, P., Smith, R. L., and Williams, R. T. (1972b) 'Molecular weight and chemical structure as factors in the biliary excretion of sulphonamides in the rat', *Xenobiotica*, **2**, 205–214.

Hirom, P. C., Hughes, R. D., and Millburn, P. (1974) 'The physiochemical factor required for biliary excretion of organic cations and anions', *Biochem. Soc. Trans.*, **2**, 327–330.

Hollingworth, R. M. (1977) 'Biochemistry and significance of transferase reactions in the metabolism of foreign chemicals', in *Handbook of Physiology Section 9, Reaction to Environmental Agents* (D. H. K. Lee, H. L. Falk, S. D. Murphy, and S. R. Geiger, Eds.), American Physiological Society, Bethesda MD, pp. 455–468.

Hsu, P., Ma, J. K. H., Jun, H. W., and Luzzi, L. A. (1974) 'Structure relationship for binding of sulfonamides and penicillins to serum albumin by fluorescence probe technique', *J. Pharm. Sci.*, **63**, 27–31.

Hutson, D. H. (1970) 'Mechanisms of biotransformation', in *Foreign Compound Metabolism in Mammals* (Senior Reporter: D. E. Hathway), The Chemical Society, London, vol. 1, pp. 314–395.

Hutson, D. H. (1972) 'Mechanisms of biotransformations', in *Foreign Compound Metabolism in Mammals* (Senior Reporter: D. E. Hathway), The Chemical Society, London, vol 2, pp. 328–397.

Hutson, D. H. (1975) 'Mechanisms of biotransformations', in *Foreign Compound Metabolism in Mammals* (Senior Reporter: D. E. Hathway), The Chemical Society, London, vol. 3, pp. 449–549.

Hutson, D. H. (1976) 'Glutathione conjugates', in *Bound and Conjugated Pesticide Residues* (D. D. Kaufman, G. G. Still, G. D. Paulson, and S. D. Bandal, Eds.), American Chem. Soc. Symposium Series, No. 29, Amer. Chem. Soc., Washington, D.C., pp. 103–131.

Irving, C. C. (1970) 'Conjugates of N-hydroxy compounds', in *Metabolic Conjugation and Metabolic Hydrolysis* (W. H. Fishman, Ed.), Academic Press, New York, vol. 1, pp. 53–119.

Irving, C. C. (1971) 'Metabolic activation of N-hydroxy compounds by conjugation', *Xenobiotica*, **1**, 387–398.

Irving, C. C. (1975) 'Comparative toxicity of N-hydroxy-2-acetylaminofluorene in several strains of rats', *Cancer Res.*, **35**, 2959–2961.

Irving, C. C. (1977) 'Influence of the aryl group on the reaction of glucuronides of aryl-acethydroxamic acids with polynucleotides', *Cancer Res.*, **37**, 524–528.

Irving, C. C. (1978) 'Reactivity of conjugates of N-hydroxylated arylamines and aryl-amides', in *Biological Oxidation of Nitrogen* (J. W. Gorrod, Ed.), Elsevier/North Holland Biomedical Press, Amsterdam, pp. 325–334.

Irving, C. C. (1979) 'Species and tissues variations in the metabolic activation of aromatic amines', in *Carcinogens: Identification and Mechanisms of Action* (A. C. Griffin and C. R. Shaw, Eds.), Raven Press, New York.

Irving, C. C. and Russel, L. T. (1970) 'Synthesis of the O-glucuronide of N-2-fluorenyl-hydroxylamine. Reaction with nucleic acids and with guanosine-5'-monophosphate', *Biochemistry*, **9**, 2471–2476.

Irving, C. C., and Wiseman, R. (1971) 'Studies on the carcinogenicity of the glucuronides of N-hydroxy-2-acetylaminofluorene and N-2-fluorenylhydroxylamine in the rat', *Cancer Res.*, **31**, 1645–1648.

James, M. O., Smith, R. L., Williams, R. T., and Reidenberg, M. (1972) 'The conjugation of phenyl acetic acid in man, sub-human primates and some non-primate species', *Proc. Roy. Soc. Ser. B.*, **182**, 25–35.

Jenner, P., and Testa, B. (1978) 'Novel pathways in drug metabolism', *Xenobiotica*, **8**, 1–25.

Kadlubar, F. F., Miller, J. A., and Miller, E. C. (1976) 'Hepatic metabolism of N-hydroxy-N-methyl-4-aminobenzene and other N-hydroxy arylamines to reactive sulfuric acid esters', *Cancer Res.*, **36**, 2350–2359.

Kadlubar, F. F., Miller, J. A., and Miller, E. C. (1977) 'Hepatic microsomal N-glu-curonidation and nucleic acid binding of N-hydroxy arylamines in relation to urinary bladder carcinogenesis', *Cancer Res.*, **37**, 805–814.

Kadlubar, F., Flammang, T., and Unruh, L. (1978) 'The role of N-hydroxy arylamine N-glucuronides in arylamine-induced urinary bladder carcinogenesis: Metabolite profiles in acidic neutral and alkaline urines of 2-naphthylamine- and 2-nitronap-thalene-treated rats', in *Conjugation in Drug Biotransformation* (A. Aitio, Ed.), Elsevier/North Holland Biomedical Press, Amsterdam, pp. 443–454.

Keberle, H., Hoffmann, K., and Bernhard, K. (1962) 'The metabolism of glutethimide (Doriden®)', *Experientia*, **18**, 105–152.

Kiese, M. (1966) 'The biochemical production of ferrihemoglobin forming derivatives from aromatic amines and mechanisms of ferrihemoglobin formation', *Pharmacol. Reviews*, **18**, 1091–1161.

King, C. M. (1974) 'Mechanisms of reaction, tissue distribution and inhibition of aryl hydroxamic acid aryltransferase', *Cancer Res.*, **34**, 1503–1515.

King, C. M., and Allaben, W. T. (1978) 'The role of aryl hydroxamic acid N-O-acyl-transferase in the carcinogenicity of aromatic amines', in *Conjugation Reactions in Drug Biotransformations* (A. Aitio, Ed.), Elsevier/North Holland Biomedical Press, Amster-dam, pp. 431–441.

King, C. M., and Olive, C. W. (1975) 'Comparative effects of strain species and sex on the acyltransferase-catalyzed activations of N-hydroxy N-2-fluorenylacetamide', *Cancer Res.*, **35**, 906–912.

King, C. M., Traub, N. R., Cardona, R. A., and Howard, R. B. (1976) 'Comparative adduct formation of 4-aminobiphenyl and 2-aminofluorene derivatives with macro-molecules of isolated liver parenchymal cells', *Cancer Res.*, **36**, 2374–2381.

King, C. M., Allaben, W. T., Lazear, E. J., Louie, S. C., and Weeks, C. E. (1978) 'Influence of the acyl group on aryl-hydroxamic acid N-O-acyltransferase—catalysed mutagenicity and metabolic activation of N-acyl-N-2-fluorenylhydroxyl amines', in *Biological Oxidation of Nitrogen* (J. W. Gorrod, Ed.), Elsevier, Amsterdam, pp. 335–340.

Kinoshita, N. and Gelboin (1978) ' β-Glucuronidase catalysed hydrolysis of benzo(a)pyrene-3-glucuronide and binding to DNA', *Science*, **199**, 307–309.

Klaassen, C. D. (1977) 'Biliary excretion', in *Handbook of Physiology, Section 9, Reactions to Environmental Agents* (D. H. K. Lee, H. L. Falk, S. D. Murphy, and S. R. Geiger, Eds.), American Physiological Society, Washington D.C., pp. 537–553.

Labella, F. S., Pinsky, C., and Havlicek, V. (1979) 'Morphine derivatives with diminished opiate receptor potency show enhanced central excitory activity', *Brain Research*, **174**, 263–271.

Lech, J. L., and Statham, C. N. (1975) 'Rate of glucuronide formation in the selective toxicity of 3-trifluoromethyl-4-nitrophenol (TFM) for the sea lamprey: comparative aspects of TFM uptake and conjugation in sea lamprey and rainbow trout', *Toxicol. Appl. Pharmacol.*, **31**, 150–158.

Levy, G. (1978) 'Pharmacokinetic and toxicologic implications of glucuronide and sulfate conjugation of certain nonnarcotic analgesics in man', in *Conjugation Reactions in Drug Biotransformation* (A. Aitio, Ed.), Elsevier, New York, pp. 469–476.

Lotlikar, P. D. (1972) 'Acylation of carcinogenic aromatic hydroxamic acids by acetyl-CoA and carbonyl phosphate to form reactive esters', in *Biological Oxidation of Nitrogen in Organic Molecules* (J. W. Bridges, J. W. Gorrod, and D. V. Parke, Eds.), Taylor and Francis Ltd., London, pp. 231–232.

McBain, J. B., and Menn, J. J. (1969) 'S-methylation, oxidation, hydroxylation and conjugation of thiophenol in the rat', *Biochem. Pharmacol.*, **18**, 2282–2284.

Marshall, T. C., and Dorough, H. W. (1979) 'Biliary excretion of carbamate insecticides in the rat', *Pest. Biochem. Physiol.*, **11**, 56–63.

Merits, I. (1976) 'Formation and metabolism of [^{14}C]dopamine 3-O-sulfate in dog, rat and guinea pig', *Biochem. Pharmacol.*, **25**, 829–833.

Miettinen, T. A., and Leskinen, E. (1970) 'Glucuronic acid pathway', in *Metabolic Conjugation and Metabolic Hydrolysis* (W. H. Fishman, Ed.), Academic Press, New York, vol. 1, pp. 157–237.

Miller, J. A. (1970) 'Carcinogenesis by chemicals: an overview', *Cancer Res.*, **3**, 559–576.

Miller, J. A., and Miller, E. C. (1969) 'Metabolic activation of carcinogenic aromatic amines and amides via *N*-hydroxylation and *N*-hydroxyesterification and its relationship to ultimate carcinogens as electrophillic reactants', in *Physiochemical Mechanisms of Carcinogenesis* (E. D. Bergman and B. Pullman, Eds.), Academic Press, New York, pp. 237–261.

Miller, E. C., and Miller, J. A. (1976) 'Hepatocarcinogenesis by chemicals' in *Progress in Liver Diseases* (H. Poppen, and F. Schaffner, Eds.), Grune and Stratton, New York, vol. V, pp. 699–711.

Miller, J. A., and Miller, E. C. (1977) 'The concept of reactive electrophilic metabolites in chemical carcinogenesis: recent results with aromatic amines, Safrole and Aflatoxin B$_1$', in *Biological Reactive Intermediates* (D. J. Jollow, J. J. Kocsis, R. Snyder, and H. Vanio, Eds.), Plenum Press, New York, pp. 6–24.

Mitchell, J. R., and Jollow, D. J. (1974) 'Metabolic activation of acetaminophen, furosemide and isoniazid to hepatotoxic substances', in *Drug Interactions* (P. L. Morsell, S. Garattini, and S. N. Cohen, Eds.), Raven Press, New York, pp. 65–79.

Mitchell, J. R., and Jollow, D. J. (1975) 'Metabolic activation of drugs to toxic substances', *Gastroenterology*, **68**, 392–410.

Mitchell, J. R., Jollow, D. J., Potter, W. Z., Gillette, J. R., and Brodie, B. B. (1973) 'Acetoaminophen-induced hepatic necrosis. IV protective role of glutathione', *J. Pharmacol. Exp. Ther.*, **187**, 211–217.

Mitchell, J. R., Nelson, S., Thorgeirsson, S. S., McMurty, R. J., and Dybing, E. (1976a). 'Metabolic activation: biochemical basis for many drug-induced liver injuries', in

Progress In Liver Diseases (H. Poppen and F. Schaffner, Eds.), Grune and Stratton, New York, pp. 259–279.

Mitchell, J. R., Zimmerman, H. J., Ishak, K. G., Thorgeirsson, U. P., Timbrell, J. A., Snodgrass, W. R., and Nelson, S. D. (1976b) 'Isoniazid liver injury: clinical spectrum, pathology and probable pathogenesis', *Annals Intern. Med.*, **84**, 181–192.

Mori, M., Oguri, K., Yoshimura, H., Shimomura, K., Kamata, O., and Veki, S. (1972) 'Chemical synthesis and analgesic effect of morphine ethreal sulfates', *Life Sci.*, **11**, 255–269.

Mulder, G. J., Hinson, J. A., and Gillette, J. R. (1977) 'Generation of reactive metabolites of *N*-hydroxy-phenacetin by glucuronidation and sulfation', *Biochem. Pharmacol.*, **26**, 189–196.

Norling, A., and Hänninen, O. (1980) 'Glucuronide and sulphate binding to subcellular fractions of rat liver', *Acta Pharmacol. Toxicol.*, **46**, 362–365.

Oguri, K. (1980) 'Conjugated metabolites of morphine and the related compounds and their pharmacological activity', *Yakugaku Zasshi*, **100**, 117–125.

Parke, D. V., and Smith, R. L., (Eds.) (1977) *Drug Metabolism—From Microbe to Man*, Taylor and Francis Ltd., London.

Paulson, G. D. (1976) 'Sulfate ester conjugates—their synthesis, purification, hydrolysis, and chemical spectral properties', in *Bound and Conjugated Pesticide Residues* (D. D. Kaufman, G. G. Still, G. D. Paulson, and S. D. Bandal, Eds.), Amer. Chem. Soc. Symposium Series, No. 29, Washington, D.C., pp. 86–102.

Paulson, G. D. (1980) 'Conjugation of foreign chemicals by animals', *Residue Reviews*, **76**, 31–72.

Peters, L. (1961) 'Urinary excretion of drugs', in *Metabolic Factors Controlling Drug Action* (B. B. Brodie, E. G. Erdös, R. P. Maickel, and P. Lingren, Eds.), Pergamon Press, Oxford, vol. 6, pp. 179–190.

Potter, W. Z., Thorgeirsson, S. S., Jollow, D. J., and Mitchell, J. R. (1974) 'Acetoaminophen-induced hepatic necrosis V, correlation of hepatic necrosis, covalent binding and glutathione depletion in hamsters', *Pharmacol.*, **12**, 129–143.

Powell, G. M., and Curtis, C. G. (1966) 'Whole-body radioautography as applied to the study of sulphate esters, *Biochem. J.*, **99**, 34–35.

Radomski, J. L., Hearn, W. L., Radomski, T., Moreno, H., and Scott, W. E. (1977) 'Isolation of the glucuronic acid conjugate of *N*-hydroxy-4-aminobiphenyl from dog urine and its mutagenic activity', *Cancer Res.*, **37**, 1757–1762.

Shaffer, J. M., and Bieter, R. N. (1950) 'Conversion of acetylsulfonamides to the unconjugated form by the chicken kidney', *J. Pharm. Exp. Ther.*, **100**, 192–200.

Shanker, L. S. (1968) 'Secretion of organic compounds in bile', in *Handbook of Physiology*, Vol. V, Section 6, American Physiological Society, Washington D.C., pp. 2433–2449.

Shimomura, K., Kamata, O., Ueki, S., Ida, S., Oguri, K., Yoshimura, H., and Tsukamoto, H. (1971) 'Analgesic effect of morphine glucuronide', *Tohoku, J. Exp. Med.*, **105**, 45–52.

Smith, R. L., and Williams, R. T. (1966) 'Implications of the conjugation of drugs and other exogenous compounds', in *Glucuronic Acid—Free and Combined* (G. J. Dutton, Ed.), Academic Press, New York, pp. 457–491.

Sperber, I. (1948) 'The excretion of some glucuronic acid derivatives and phenol sulphuric esters in the chicken', *Ann. Royal Ag. College of Sweden*, **15**, 317–349.

Thor, H., Moldeus, N. D., and Orrenius, S. (1979) 'Isolated liver cells for the study of drug toxicity', *Symp. Med. Hoechst.*, **14**, 355–371.

Thor, H., and Orrenius, S. (1980) 'The mechanism of bromobenzene-induced cytotoxicity studied with isolated hepatocytes', *Arch. Toxicol.*, **44**, 31–43.

Timbrell, J. A. (1979) 'The role of metabolism in the hepatotoxicity of isoniazid and iproniazid', *Drug Met. Rev.*, **10**, 125–147.

Uesugi, T., Ikeda, M., and Kanei, Y. (1974) 'Studies on the biliary excretion mechanisms of drugs III active transport of glucuronides into bile in rats', *Ceym. Pharm. Bull.*, **22**, 433–438.

van Bladeren, P. J., van der Gen, A., Breimer, D. D., and Mohn, G. R. (1979) 'Stereoselective activation of vicinal dihalogen compounds to mutagens by glutathione conjugation', *Biochem. Pharmacol.*, **28**, 2521–2524.

Vina, J., Estrela, J. M., Guerri, C., and Romero, F. J. (1980) 'Effect of ethanol on glutathione concentration in isolated hepatocytes', *Biochem. J.*, **188**, 549–552.

Walle, T., Conradi, E. C., Walle, U. K., Fagan, T. C., and Gaffney, T. C. (1979a) 'Propanolol glucuronide cumulation during long-term propranolol therapy: A proposed storage mechanisms for propranolol', *Clin. Pharmacol. Ther.*, **26**, 686–695.

Walle, T., Fagan, T. C., Conradi, E. C., Walle, U. K., and Gaffney, T. C. (1979b) 'Presystematic and systematic glucuronidation of propranolol', *Clin. Pharmacol. Ther.*, **26**, 167–172.

Weber, W. W. (1971) 'Acetylating, deacetylating and amino acid conjugating enzymes', in *Handbook of Experimental Pharmacology. Concepts in Biochemical Pharmacology II* (B. B. Brodie and J. R. Gillette, Eds.), Springer-Verlag, New York, vol. 28, pp. 564–583.

Weisburger, J. H., Yamamoto, R. S., Williams, G. M., Grantham, P. H., Matsushima, T., and Weisburger, E. K. (1972) 'On the sulfate ester of N-hydroxy-N-1-fluorenylacetamide as a key ultimate hepatocarcinogen in the rat', *Cancer Res.*, **32**, 491–499.

Weisburger, J. H., and Weisburger, E. K. (1973) 'Biochemical formation and pharmacological, toxilogical and pathological properties of hydroxylamines and hydroxamic acids', *Pharmacological Reviews*, **25**, 1–66.

Williams, R. T. (1947) *Detoxication Mechanisms*, 1st edn., Chapman and Hall, London.

Williams, R. T. (1959) *Detoxication Mechanisms*, 2nd edn., Chapman and Hall, London.

Williams, R. T. (1974) 'Interspecies variations in the metabolism of xenobiotics', *Biochem. Soc. Trans.*, **2**, 359–377.

Williams, R. T. (1977) 'Introduction to the concept of reactive intermediates', in *Biological Reactive Intermediates* (D. J. Jollow, J. J. Kocsis, R. Snyder, and H. Vanio, Eds.), Plenum Press, New York, pp. 3–5.

Williams, R. T., and Millburn, P. (1975) 'Detoxification mechanisms—the biochemistry of foreign compounds', *Physiol. Pharmacol. Biochem.*, **12**, 211–266.

Wislocki, P. G., Borchert, P., Miller, J. A., and Miller, E. C. (1976) 'The metabolic activation of the carcinogen 1'-hydroxysafrole *in vivo* and *in vitro* and the electrophilic reactivities of possible ultimate carcinogens', *Cancer Res.*, **36**, 1686–1695.

Wood, J. L. (1970) 'Biochemistry of mercapturic acid formation', in *Metabolic Conjugation and Metabolic Hydrolysis* (W. H. Fishman, Ed.), Academic Press, New York, vol. 2, pp. 261–298.

Yoshimura, H., Idans, O., Oguri, K., and Tsukamoto, H. (1972) 'Biochemical basis for analgesic activity of morphine-6-glucuronide-I: Penetration of morphine-6-glucuronide in the brain of rats', *Biochem. Pharmacol.*, **22**, 1423–1430.

Young, L. (1977) 'The metabolism of foreign compounds—history and development', in *Drug Metabolism—From Microbe to Man* (D. V. Parke and R. L. Smith, Eds.), Taylor and Francis Ltd., London, pp. 1–11.

Subject Index

Contents—Volume 1

Contents—Volume 2